Praise for *The Way of the Water Priestess*

"Calling all water witches and merfolk! Annwyn has created a sacred text to bring you home to this most powerful and necessary element. *The Way of the Water Priestess* gives empowering and magical ways to work with our mother, the waters. Whether well-versed in the ways of water or just dipping your toes into water magic, you will love this thoroughly researched and engaging book."

—Madame Pamita, author of
Madame Pamita's Magical Tarot

"If you've ever gazed at a body of water and felt its healing, energizing power, then you've found a lifelong treasure in *The Way of the Water Priestess*. Just reading it is a spiritual journey. With every page, you'll find yourself spiraling deeper into Annwyn's sacred watery realm."

—Tess Whitehurst, author of *You Are Magical*

"With *The Way of the Water Priestess*, Annwyn takes us through the mists to connect with our own inner water priestess and the water goddesses."

—Amy Blackthorn, author of
Blackthorn's Botanical Brews

"In *The Way of the Water Priestess*, Annwyn pours forth a bountiful offering of beautiful, simple, clear wisdom and ritual that anyone answering the siren call of the ancient waters can hear and heed. Every water priestess can find fresh inspiration and cleansing guidance within this book."

—Laurelei Black, aut

THE WAY OF THE
WATER
PRIESTESS

*Entering the World
of Water Magic*

ANNWYN AVALON

WEISER
BOOKS

This edition first published in 2024 by Weiser Books, an imprint of
Red Wheel/Weiser, LLC
With offices at:
65 Parker Street, Suite 7
Newburyport, MA 01950
www.redwheelweiser.com

ISBN: 978-1-57863-724-9
Library of Congress Cataloging-in-Publication Data available upon request.

Cover design by Kathryn Sky-Peck
Cover image adapted from Circe Invidiosa (1892) by John William Waterhouse
Interior illustrations by Opia Designs
Interior by Kasandra Cook
Typeset in Incognito

Printed in the United States of America
IBI

10 9 8 7 6 5 4 3 2 1

To Sulis, Melusine, Morgan, and the water spirits.

CONTENTS

CONTENTS

INTRODUCTION

Why is it that we are so drawn to the sea? Why have there been so many temples erected in the name of water goddesses and so many groves named for watery nymphs? Why do we continuously connect the ocean with femininity and water with life? Could it be a distant memory–part of our collective consciousness that attunes us to the spirit of water as she calls out to her daughters to remember the ancient ways? To remember what it was like when we held the Earth as sacred and worshiped the great goddesses at sacred wells?

So many of us are called to the water; we are drawn to her shores, where we find healing and solace. With the rise of toxic masculinity, many of these gentle but powerful practices were lost to time. Fortunately, they have been documented across various cultures throughout history, hidden within the pages of ancient texts and painted and carved on ancient temple ruins. Numerous depictions have survived of sacred women tending water, bearing water, cleansing through water, and performing religious rites in conjunction with sacred water places.

In fact, examples of priestesses and sacred women associated with water are legion and can be found in every corner of the world in every time. And it is safe to assume that they will survive in times to come. Some popular examples include Rome's Vestal Virgins, who were tasked with collecting sacred waters; the Pythia, who bathed in sacred springs before engaging in oracular trance work; the well maidens and priestesses of Celtic lore, some of whose practices were documented in letters written by Roman generals and sent back to Caesar. Some of the most detailed accounts of Gallic and Breton traditions were preserved in this correspondence.

During Christianity's rise, women like St. Keynes, St. Hilda, and St. Brigit also practiced sacred and miraculous water arts. While the Church demonized women and contributed to the demise of ancient temples and priestess practices, they could not deny their existence or stamp them out completely. In fact, during the medieval era, many Christian churches were erected on sites that were once home to ancient Pagan temples and other places associated with springs, holy wells, and sacred landscapes. This is why we find so many associations with water in old churches and cathedrals, like the carving of a mermaid on a bench in a chapel in St. Senara's church in the village of Zennor in Cornwall that dates from the 1400s. No doubt these remnants of ancient water practices were preserved *sub rosa* (hidden) in carvings, motifs, and folklore in the buildings and practices of the new religion.

These ancient magical water mysteries were practiced by Pagan priestesses, who held them as sacred and occult. While medieval Christianity did its best to suppress these ancestral practices, they survived in local lore in some of the healing and spiritual practices often associated with witchcraft. In fact, as priestesses were suppressed and their practices condemned, we begin to see the same types of women described as witches.

A perfect example of this is found in the demonization of Morgan la Fae, or Morgana. Morgan is deeply connected to Avalon and her role as a priestess of Avalon can give us insight into these practices. Today, Morgan's spirit can be felt in Glastonbury, the modern-day Avalon, calling sacred women to tend the waters once more. In her original description from the 12th-century text the *Vita Merlini* (1150 CE), Morgan is described as having knowledge of herbs and healing ointments, the ability to shapeshift, and the gift of second sight. It was also recorded that she taught her sisters mathematics and had the ability to manifest "new wings like Daedalus," who, not born with wings, rather fashioned them out of wax. This ability to appear with wings or without may be linked to her shapeshifting ability, or it could indicate her role as a faery queen. While there is no other information about this, it is interesting that Morgan is described as having wings that may be artificial, or that she has the ability to remove them at will.

Years later, Morgan emerges again in a new guise as an evil sorceress. In the *Vulgate Cycle* of the Arthurian romances, which was written in the early 13th century, she becomes a witch who is described as the evil half-sister of King Arthur. This tradition of equating strange healing women with witches went on for hundreds of years. Finally, however, we are beginning to see these priestesses re-emerge as wise women who still work these types of magic in their communities, but under different names–among them Initiatory Wicca, Metaphysics, Dion Fortune's Community of Inner Light, and so many more. It is here that we find modern-day magical practitioners and the powers of the sacred feminine returning to heal the land and restore sovereignty to the waters once again.

Both these historical survivals and these modern practices provide us with clues to the rites of ancestral water priestesses that can now be pieced back together to create a path for modern water priestess that is rooted in the lore of our ancient mothers. In the pages of this book, you will explore these ancient traditions, rediscover the sacredness of water, and learn how to incorporate these powers into your own modern water practice. You will work with ancestral priestesses through meditation and vision journeys to gain wisdom from the Otherworld and acquaint yourself with how they lived and worked as they followed their ancient water paths.

In the first three chapters, you will learn who these women were and explore some of their practices. In later chapters, you will dive deeper into the priestess arts and learn how to implement them as a part of your own path. As water priestesses, we can make a change in our world—before the water wars begin, before only the elite have access to the waters. It is time to rise up and reclaim the ancient arts.

Chapter One

RETURN TO ANCIENT WISDOM

*T*he word "priestess" is a noun that describes a woman who performs the sacred rites of a particular religion. It is the feminine variant of "priest." A priestess is thus a female officiant of sacred acts and a facilitator of ritual who serves a particular religion or deity(s), especially of a non-Christian religion. In history, priestesses were associated with the worship and temple-tending of either male or female deities. But the roles embodied by the word "priestess" are vast, ever-evolving, and incapable of being limited by mere definitions.

When you view the word "priestess" as describing an active participant in sacred activities, you acquire a broader understand of a priestess' role. As an active participant in sacred rites, a priestess may be a ritual facilitator or guide, a healer or spell-caster, a prophetess or seer. A priestess defines herself by the role she chooses to fulfill and the way she exercises her powers. In fact, she may not even choose to call herself a priestess, but prefer rather to be known by the tangible fruits of her magical endeavors.

In the past there were magical schools that taught priestesses the sacred arts. After a period of study, novices were ritually initiated into the school or temple as priestesses, usually by other practitioners acting on behalf of the goddess or spirit they served. These newly initiated priestesses then either continued on to advanced studies or began their work in service to the temple or their particular path. Some modern schools are exclusive to women, while others initiate both male and female novices.

It is also important to know that there is no such thing as a high priestess in the tradition, and certainly no one should self-aggrandize themselves by taking on this title, as it would be a disgrace to those who have gone before and a grievous misrepresentation of their skill and work. The term "high priestess" is usually applied only to someone in a tradition or temple who has passed through several levels of initiations and elevations. It is used to denote their role within the tradition and the temple, and as an indication of how many students, circles, rituals, and temple duties they preside over. It may also refer to a seasonal or temporary position, depending on the tradition and its ritual structure. This path is about service, not self-elevation. Always be cautious of this kind of self-aggrandizement. There are many priestesses in the world, many of whom are good teachers who work from a place of service, not self-gain.

The way of the water priestess is a Pagan path, as you will find in the pages that follow. We have undeniable proof that there were ancient practices that drew on the sacredness of water and service to water goddesses. We know of priestesses past and present who have tended water temples. Their practices may have been obscured by time, but they were not lost. And today, they are being revived.

It is hard to nail down when this revival began and who started it, as many different groups working for similar goals began around the same time. Doubtless New Age, Metaphysical, Pagan, and Wiccan practices helped redevelop these ancient arts and brought them into the modern age. Wiccan and Pagan revivals brought about a surge of interest in ancient practices and ways of living. As these traditions gained ground, circles, groves, and covens grew and expanded, until, today, the freedom to practice these occult and Pagan arts without fear of being killed or jailed has been reborn. While we still have far to go, we no longer have to hide in the shadows, afraid for our very lives or for our way of life. We are now free to practice when and how we wish. Indeed, many of us are inexplicably pulled to these arts, perhaps by past-life memories or perhaps by the desperate cry of the water to return to the temple. Perhaps we are drawn by the need to heal and protect the waters, or to teach its wisdom and mysteries. Whatever the reasons, priestesses are returning to the sacred water, the sacred land, and the sacred life.

Like the ancient priestesses from whom we draw wisdom, we are also called to worship the pure raw beauty that is water—its ever-changing form and its many guises as the great water goddess, the Oceanids, and the nymphs who reside in the sacred founts. The powerful water goddesses pull at our heart strings and urge us to remember what we forgot when we were sleeping. They are gently washing the mud from our eyes and reminding us that there was once a time when the water was worshiped, when it was revered as a life-giving force, when women dedicated their entire lives to walk the way of the priestess and enter the water temple.

The Role of a Water Priestess

The role of a water priestess is one dedicated to the service of the water and water spirits. This is not a new or isolated practice. In fact, it is quite ancient and is cross-cultural. Since antiquity, and possibly since the beginning of time, women have tended sacred waters. These sacred sources were often found near temples or in groves, or perhaps were just local water sources. For hundreds of generations, these arts were passed on from culture to culture, eventually being forced almost completely underground by the advent of organized religions. Modern water priestesses are often drawn to serve a particular water goddess. Some are directly called by a particular body of water or spirit to be the human conduit for it. Priestesses who don't work directly with water spirits or goddesses sometimes choose instead to work with a particular type of water, like the sea. Sometimes they call themselves sea priestesses. Others prefer to be called healers, or perhaps water healers or water magicians. Some water priestesses work through the oracular arts and may refer to themselves by ancient titles like *Pythia* (Greek) or *Volva* (Norse). Still others walk the crooked path of the magical arts as water witches, enchantresses, or sorceresses.

No matter what *you* choose to call yourself as you work with water spirits, enact devotional rituals to a water goddess, or practice healing water magic, you are doing the work of the water priestess. Whether you are the water bearer holding the sacred waters of your lands, healing the community and protecting the water and its creatures, or serving the watery will of an ancient powerful deity–these are all acceptable ways to walk the way of the water priestess.

The work of a water priestess is expressed in various sacred practices like enchanting the waters, facilitating rituals, creating healing ceremonies, and preparing sacred baths. Water priestesses offer devotional practices to the water or water spirits, commune with water spirits, cleanse and purify the waters, and perform healing rituals with sacred water. As a water priestess, you may find yourself called to use water to heal your community, or you may be called to heal the water through energy work. You may be called to protect the water from more harm with magical or mundane water activism. You may feel led to join beach or river clean-ups, or to bless water for the land or your community. Your mission may be to revive the old temple arts in your community, or to find other ways to honor the water with singing, dancing, drumming, or chanting at the water's edge. We will explore more of these practices in the pages that follow.

While many priestesses are actively devoted to a higher power and facilitate religious or spiritual rites, many choose to focus their work on their communities. People who choose this path usually spend several years apprenticed to a trained practitioner or in a training program to reach the rank and experience of a priestess. Only through diligence and hard work will you become exactly the person you wish to be. Constant training, constant learning, and constant shadow work will help you to become a powerful priestess–always balancing, always healing, and always seeking out knowledge and new ways to look at the world.

Embarking on the Path

There are three different portals through which you can pass when you embark upon the path of the water priestess: dedication, ritual initiation, and spirit initiation. While you cannot initiate yourself as a priestess, you can dedicate your life to the sacred waters. Dedication rituals are incredibly powerful. They usually consist of a solitary ceremony performed at your sacred water space or home altar. They mark a moment on your journey when you relinquish control and surrender to the sacredness of water. It is here that you say your vows and dedicate yourself to service. This act may start the process of ritual or spirit initiation or encourage you to seek it's path. You will find a dedication ritual at the end of this book.

Dedication rituals can be quite beautiful. But initiations are something quite different, with a history and tradition all their own. Ritual initiation recognizes a priestess' achievements and marks her acceptance into the temple and into the service of the goddess. While this formal training is not necessary to become a priestess, the hard work and dedication it reflects should be greatly respected. In many cases, the training serves as a kind of peer review that gives initiates the backing of others from that particular school or temple, who can then vouch for their knowledge. Through the training process, novices prove that they are ready to become full-fledged practitioners and that they should be respected as such. When you see the term "initiated priestess," it indicates someone who has been formally trained and who has studied for at least one year–in many cases more. Priestesses who have passed through ritual initiation have the respect of their school, their temple, and their peers.

Ritual initiation flows from ancient tradition in which young maidens were selected to join the priestess path and then brought through a sacred rite. With the rise of Wicca, many priestesses have passed through ritual initiations that solidify their role in both the mundane and spiritual realms. These consist of elaborate ceremonies in which they are celebrated, challenged to show their skills, or required to demonstrate that the spirits will recognize and accept them as someone sacred to the tradition. Many believe that you cannot truly be a priestess if the spirits have not called you. There is some truth to this. I believe, however, that you can surely do the work of a priestess diligently and humbly until the spirits take notice and recognize you as their priestess.

If you choose to follow the path of initiation, at no time should you be coerced into being naked or having sex with any circle or temple member. Although it is true that some traditions follow these practices, you do not have to participate in them if you are not comfortable with them, and you should never be pressured to do so. Not engaging in these practices will in no way hinder your progress on the path. Rather, your choice not to participate is evidence of your own sovereignty, and this will carry far greater weight than compromising yourself to fulfill requirements.

Spirit initiation is probably the least glamorous type of initiation. In fact, it can be quite messy, and usually involves a long and drawn-out process. Spirit initiations cannot be summed up in one paragraph or even one book. In fact, each person who has experienced this type of initiation could probably write a volume on the trials and tribulations they experienced during the process. True spirit initiation is often fraught with trauma, and most people who have undergone it don't tend to brag about it or use it as a marketing ploy. Those who have gone through facilitated ritual initiation tend to view it as a mark of positive achievement within a group, a community, a coven, or a circle, using it denote a certain amount of skill and experience. But spirit initiation is so intimate and so intense that those who go through it generally view it as one of the most traumatic and personal experiences of their lives and don't tend to share that experience freely.

Some call spirit initiation the path of death and rebirth, the path of transformation, the dark night of the soul, or the path of the wounded healer. All of these terms describe a process in which a person undergoes a radical transformation akin to that signified by the Tower card found in the Rider-Waite-Smith tarot deck, as well as in many others. This is not the same as the change brought about by a ritual initiation, although that may take place before this process happens, directly after it, or not at all. In this case, initiates are put through a drastic spirit journey in which many things are stripped away from them. This can often happen before they take on the role of priestess as well. In fact, many who go through this process and survive emerge so thoroughly changed that they take different names, modify their appearance, or change their personalities completely, having triumphed over shadow and the forces that held them back.

Spirit initiation is based in healing—in the transformation of pain into something positive that can be used to heal or bring beauty to the world. This often takes many years and usually leaves the person completely changed forever. We see this in Greek myths like the stories of Chiron, the wounded healer, and Persephone and her journey into the underworld. Chiron was a centaur who was wounded in battle and took shelter in a cave, where he fell prey to depression, anger, and many other feelings associated with his trauma. Through diligence and perseverance, he eventually healed himself and became the archetype of the wounded healer, learning from his wounds and using that knowledge to heal others.

Trauma comes in many forms. And Chiron teaches us that, no matter how much pain you go through, no matter where you are on your path, you can find healing and use that gift to heal others. This is essential work for the priestess. Many of us have survived physical trauma, childhood trauma, or psychological and even spiritual trauma. This may come in the form of near-death experiences, tragic accidents, sexual assault, or other painful events. These experiences can take you on your own personal underworld journey, like that of Persephone, who knew she must walk in darkness until it was time to come back to the light. There may be periods in your life that seem like an underworld journey that is gut-wrenchingly difficult to navigate. You may even think that you have healed a portion of your trauma, only to have it resurface in a powerful and destructive way. Indeed, you may find that you can never heal completely, and that regular healing rituals are necessary to maintain your balance.

Those who have undergone this type of underworld journey know that good and evil exist in every extreme. You may find the exact healing you need in the depths of the darkness; your shadow may be an ugly sea hag who scares the crap out of you, but brings powerful messages to the surface. Please remember, this is still a part of you and you are whole even in your most broken state. You may pass through the gates of death into your own personal hell, but the journey gives you wisdom beyond what others may see. Those who are strong enough to fight for their own lives and balance often come out on the other side of these experiences completely changed and ready to take on the mantle of a water priestess. And by no means do you even have to take this journey. There are other ways to walk the priestess path. This is just the long way around.

Spirit initiation itself does not make you a priestess. Instead, the process of journeying through the underworld and coming out on the other side gives you knowledge and wisdom. It teaches you how to be strong, how to fight for your own life—and then, eventually, for others and for the water. In spirit initiation, you pass through a type of death in which your shadow struggles to find balance with your conscious self. It is as if a potter takes formless and lumpy wet clay and, through constant reshaping, kneading, and carving, transforms it into a work of art. So also is the priestess shaped and reborn through the death-and-rebirth cycle, molded like the clay, shaped, carved away, and changed.

I do not recommend this path, but if you have already traveled or are currently on it, you can be encouraged by the fact that many others have taken this treacherous journey before and been transformed into powerful priestesses who are ready and willing to share their healing and gifts with others.

The Power of Healing

A water priestess learns from her own struggles and the struggles of those around her that she can bring about great healing and forgiveness, as well as great love and transformation. In her work, she finds balance between the light and the dark. When she feels herself leaning too far in one direction, she seeks balance. She rides the hedge, listening to her guides, goddesses, and spirits. She is a devotee of her path and of those she serves. She has been to the underworld; she has looked death and life's challenges in the face and survived. Because of these experiences, she is able to guide others through their own dark night of the soul, their journeys through their own personal hells. She does not force healing, but is there when healing is needed. She knows that, although there is great power in speech, there is also great power in keeping silent. She holds her tongue (when she can) and tries to speak only wisdom. She also knows she is human and fallible, and so constantly seeks healing for herself. She knows that true power comes from finding the wisdom within balance and seeks this wisdom to better herself in order to better serve others.

The first step to healing pain and trauma is to acknowledge their presence in your life and recognize that they are a part of you. Of course, you don't have to stay trapped in these experiences. But, in order to heal, you must first recognize and acknowledge them. Once that is done, the healing process can begin. This is not a short process. In fact, it can be a very long and drawn-out one that is comprised of many layers.

If you find yourself at the beginning of this journey, or are struggling and feel isolated, please know that you do not have to go through this healing process alone. There are many others out there with similar stories. There are professionals who can help you and councilors who can listen. For many, this process is necessary for them to become their best selves, to heal what others have done to them, and to share those healing gifts with others.

Although the healing path may be lonely and hard, for some reason we are driven to this work–perhaps because we want to serve, or perhaps because we want to heal our own wounds.

Through that healing, we find our own power to guide others on a similar journey. This journey takes you through both the mundane and the spirit worlds, and to the depths of your soul, your shadow, and the underworld.

A final note here. While you may feel very comfortable on your own water path, remember that a priestess is first and foremost a facilitator–an advocate for the water and a conduit for its spirits. You hold within your hands the souls of seekers and those who are lost. If you are not constantly working on yourself–becoming more educated and evolving as a priestess– you risk the lives, souls, and well-being of those who come to your temple for help, or for healing, or simply to work with the water. The way of the water priestess is not a path of arrogance, but one of humble devotion and powerful change.

Exercise: Understanding Death and Rebirth

Spend time in meditation and later with your Book of Tides (see page 44). What types of experiences have you had that follow the death-and-rebirth cycle? What type of energy did you encounter? Were you able to heal it or are you still working on it today? If you don't have any connection to this energy, that is okay. Spend time meditating on gratitude for not having to learn this lesson in this lifetime. I truly do not wish this process on anyone. And I admire those who have undergone it and become better for it.

Exercise: Connecting with Water

The way of the water priestess is a way of life that puts you in constant connection with the water and with your desire to serve your community. In order for you truly to walk, or perhaps swim, this path, water must become a focus of your daily routine. This can include activities like water rituals and sacred bathing, or even practices as simple as drinking water every morning.

Start every day by consciously drinking water. Taking in water first thing in the morning helps get your body flowing and hydrates you after your rest. Before you begin to sip your morning water, try putting it in a glass cup and using sound resonance to charge it with beauty and energy. You can do this by simply singing to the water and using the tone of your own voice. If you prefer, you can play a sacred song or use a singing bowl or bell to charge the water with high vibrations. This is a great way to align your body with your purpose and charge it with sacred water.

Note how the water makes you feel and how your day goes after you have positively charged it. If you don't have a singing bowl or a way to play music to the water, try using your voice to intone the sacred sounds of vowels to enchant it. First, bring yourself to a calm and centered state. Then, holding your glass, pick one vowel for the day. Vibrate the vowel with your voice into the top of the glass. This will help empower your day by connecting you more intimately to the water and attuning you to it.

Your Book of Tides

It is really important that you record all of your spiritual work in a Book of Tides. You will find over time that this journal is your best teacher and a powerful magical tool. This should be a place where you keep a record of your most intimate moments and your goals as a priestess–a place where you store your meditations, your oracular readings, your rituals, and the contacts you have with water spirits.

Your Book of Tides will become your teacher as you progress along your path. In it, record both your successes and your failures, so you have the opportunity to review them and implement what you learned in your next ritual or meditation. It can also be a resource as you begin to create your own rituals and to formulate your own workings.

When you begin to do oracular work, the practice of recording every reading and reviewing them later will show you, over time, exactly where you are successful, the ways and times at which you were successful, what in your practice needs work, or simply which techniques do not suit you. You may also find that you gravitate to a specific practice more than others.

Exercise: Learning the Lessons of Water

Water has many lessons to teach us. Over time, these lessons, like water, will flow through your life and change you. Sometimes you may be greeted with the lessons of calm waters; at other times, that of treacherous waters. Sometimes the lesson may be to play the game slowly, as with the water that formed the Grand Canyon; at other times, it may be to go with the flow. Sometimes the message will be harsher and will teach about rebirth and the destruction of death; sometimes it will be more hopeful and will speak of transformation and healing power. And sometimes, water will just remind you of how incredibly powerful you are.

This exercise will help you begin to attune to the frequency of water and connect with its energy. You can do this meditation in your temple room, by your altar, in the bath, or at the beach. In fact, I encourage you to do it often and with different types of water. You will find that the lesson of the river is different from the lesson of the ocean, which is different still from the lesson that floral water, a sacred bath, or a gem elixir may teach you.

You will need a clear glass cup, bowl, or vase–something that can hold water without adding color or metal to the equation. I recommend clear glass because it lets you observe the water from all sides, with only the glass separating you from it. Later, you may want to experiment with different colors of glass to see if you have different experiences. This may be useful for water rituals in which color therapy plays a role, or in creating waters to align with different energy points in the body.

Once you have selected your glass vessel, find a quiet place to meditate. The first time you do this, I recommend simply positioning yourself in front of your altar or shrine space. After you have successfully connected with the water, you can take it deeper and do this same meditation while sitting by the ocean, with your feet in a river, or in a candlelit bath.

Next, select your water source. There are many options for you here. You can simply start with water from your kitchen sink, or you may choose to connect with sacred water. If that is the case, acquire water from a sacred site or fresh-water spring. If you decide to buy bottled water, be sure to recycle the container and also consider what type of water it is. Is it distilled? Has it been processed? Was it bottled at the source, where no hands touched it? These factors will all make a difference in the messages and energy you receive. Once you have decided on the type of water you want to use, pour it into your vessel and prepare for the meditation.

Begin by finding a safe and comfortable position. It is important that you have no other distractions, like noise or body pain. Use pillows and blankets if necessary and settle down for a quiet talk with the water.

Hold the water in both hands and breathe over it, mingling your breath with its surface. Pause for a few breaths, then swirl your index finger around in the water three times in a sunwise direction. Anoint your heart center, lips, and Third Eye with the water as you begin to breathe slowly and enter into a light meditative state.

In your mind's eye, let images of water come and go. See a large angry ocean and a peaceful spring stream; see the beauty of a sun shower and the fierceness of a blizzard. Allow these images to come and go until you feel yourself settle into one to which you feel drawn. Allow the images to flow organically. Let the water communicate with you; let it speak in words, feelings, and images. Take your time and allow the water to flow through you.

When you feel the images begin to fade and the voices go quiet, take three deep breaths and slowly open your eyes. Using the same finger, swirl the water in a counter-sunwise direction and anoint your Third Eye, your lips, and your heart center. Thank the water for the connection you have experienced. If you know without a shadow of a doubt that your water is drinkable, drink the rest of it.

When you record your experience in your Book of Tides, try to answer these questions:

- What did I see?
- What did I hear?
- How did the water feel?
- What sensations did I experience?
- Was there a clear message for me?
- How can I apply these lessons to my own priestess practice and water magic?
- What was the most important part of this meditation?

Exercise: Daily Water Practice

To being walking the way of the water priestess, you must first find a starting point. If you haven't already, begin by considering what a daily water practice may look like to you. What are your particular skills? What are your unique talents? What do you love doing? Answering these questions can help you learn more about yourself and the particular path you will follow as a water priestess. There is no reason to reinvent the wheel, or to swim upstream. Start where you are and with what you have, and build from there.

In my own daily practice, I engage in consciously drinking water. While the usual recommendation is eight glasses of water a day, for the purpose of a water priestess, try aligning your intake with the number nine. You will find, in the following chapters, how this number has played a large role in rituals associated with water throughout history and across many cultures.

You may also wish to perform daily devotions to a particular goddess, a water spirit, an ancestral spirit, or just to the spirit of water itself. We will explore these options in later chapters. If none of these resonate for you, try choosing a local water source or river and visiting it each morning to perform your prayers or energy work. In chapter 4, you will learn how to set up a temple space. If you don't want to dedicate it to a spirit, consider dedicating it to a lake, a beach, a waterfall, or a river near your house. You can keep a bottle of water from this source on your altar in place of a statue or image. Of course, you can do all three as well! The important thing is to spend time in contemplation, perhaps even divination, to find out with whom and where you should work.

Remember that this is intended to be a *daily* devotional practice, and you must be sure you practice regularly. This may mean that you need to keep it short and simple. You don't have to perform an elaborate ritual to connect with the waters on a daily basis. You can just start by blessing your first glass of water each day and spending ten minutes in meditation.

Once you have established your daily ritual, you can commit to weekly rituals at your favorite body of water, or elaborate water-healing rites in your temple on every Full Moon. Whatever you choose, be sure to be consistent and to fulfill your obligations, promises, and vows.

Chapter 2

ANCESTRAL WATER PRIESTESSES

*T*he word "priestess" denotes a particular type of person engaging in a particular type of activity–a sacred woman with a sacred purpose who enacts rituals and rites in the name of something sacred. Even if a woman is not referred to as a priestess, if she is indeed engaged in these acts, she becomes a sacred vessel–and thus, a priestess.

We know very little about ancient priestesses, but, over time, through archeology, we have been able to piece together a little about their lives. The first priestess to appear in the historical record was Enheduanna, a Sumerian high priestess who lived 4,300 years ago. She was one of the earliest known authors, having written many poems in cuneiform. Another ancient priestess comes to us from ancient Greece–Myrrhine, who, according to funerary inscriptions, was the first priestess of Athena Nike.

(She is not to be confused with the character in Aristophanes's play, *Lysistrata*.) I find it an interesting correlation that the resin often burned in sacred ceremonies and as incense in practices associated with water is myrrh.

Throughout the Greek and Roman worlds, springs were well known as dwelling places for nymphs. The Temples of Hera, Aphrodite, and Apollo at Delphi all had sacred springs associated with them. Hera immersed herself each year in the spring of Kanathos to regain her purity. Even the Vestal Virgins, who were well known as flame tenders, had the duty of drawing water from the sacred springs.

The oracle at Delphi, who was called the Pythia, was a seer whose oracular work and worship of Apollo brings light to some water practices in temples dedicated to solar deities. In fact, we find similar practices in the Roman, Celtic, and Gallic cultures. Pythia, named for the python that Apollo killed, bathed in the Castalian springs to prepare herself through cleansing and purification before her prophecies. She is also connected to the sacred serpent, a motif often associated with sacred springs, water, oracular work, and priestess arts. The cult of Apollo, although a solar cult, was further connected with water. His temple at Delphi was oriented so that the northern door faced the helical rising of the constellation Delphinus, the Dolphin, while his temple at Delos was oriented to the setting of the same constellation. Some scholars believe that it was the rising of this constellation, which occurred during the month of Bysios in the ancient Greek calendar, that let the people know the oracle was ready to be consulted. (For more on this, see *Journal of Astronomical History and Heritage* at *academia.edu*.)

In one version of the story, the muses were born after Zeus and Mnemosyne slept together for nine consecutive nights. In another, Pegasus struck his hoof upon the ground and a sacred spring or fount sprang forth. From this fount, the nine muses were born. Several accounts describe nine sisters, nine healers, and nine priestesses, indicating that the number nine was sacred in practices that involved priestesses and water. It is possible, however, that these were three different groups of women, from different times and different places, all displaying similarities and slight differences.

Germanic and Nordic Traditions

Oracular work is central to many priestess traditions and this practice has survived as the role of sacred women evolved. The lines between priestesses, sorceresses, oracles, and witches were often blurred in history, so we must consider all these practices when researching the role of water priestesses.

In Norse mythology, groups of nine were associated with the feminine and with water. Aegir and Ran, giants that lived beneath the ocean, had nine daughters. The nine mill mothers, who were the mother of Heimdall, are another example of a group of nine feminine figures in the Norse mythos. The world tree, which holds nine worlds, is central to Norse belief as well. The Valkyries were often mentioned in groups of nine, while nine wood witches appear in verse. In Germanic tribes, a seer named Veleda, who was later venerated as a goddess, lived in a tower near a river. The Norse called this sacred oracular woman Volva, and a coven of nine Volvas who had oracular powers are found in Nordic literature.

Gallic and Celtic Traditions

There are many references to water priestesses in the Gallic and Celtic worlds as well, and many of the surviving records depict these priestesses in groups of nine. The Celtic world in particular made many correlations between the number nine and water, including the nine maidens of Annwn who kindled the fire under the pearl-rimmed cauldron, the cauldron of rebirth. There are nine hazel trees around the Well of Wisdom, as well as nine waves for healing, nine sacred woods for sacred fires, and nine herbs. Nine flowers form Bloudwedd. The nine witches of Gloucester in Welsh lore may also be a part of this tradition.

Another interesting correlation with a group of nine maiden comes from Cornwall, where legends survive of nine sisters being buried in a stone circle. However, later legends tell us that these stones are, in fact, nine wicked sisters who were turned to stone because of their evil ways. This legend, like so many others, tends to demonize sacred women, priestesses, and the feminine arts. The nine witches of Gloucester may be another example of this.

According to folklore from the Celtic faery tradition, the Korrigan also appeared as a group of nine women. They are sometimes described as robed in white and dancing around a fount by the light of the Full Moon. They also appear near Samhain, lurking near standing stones. They are often described as dwarf-like faeries, but also as sirens with long gold or white hair, pale skin, and red eyes. They are usually seen around dusk, as they are only beautiful at night. During the day, they appear with wrinkled hag-like flesh.

Like mermaids and sirens, the Korrigan are thought to steal souls and to lure men to their watery graves. They are also described as Druidic priestesses who stood in opposition to the new Christian religion and were said to hate anything to do with the Church—especially the Virgin Mary and priests. They have the ability to shapeshift, to foretell the future, and to move with incredible speed. They have also been associated with sacred founts and wells.

The Isle of Sena boasted a group of nine priestesses, as did the Isle of Avalon. In Celtic lore, the maidens of Glen Ogilvy, in Scotland, also numbered nine. Though they are often referred to as sisters, the legends imply more. One ballad that is associated with them specifically states that St. Donivald's nine daughters built a temple in the glen. After their father's death, they left the glen to live in Abernethy until their death, after which they were buried under an oak tree, a tree sacred to the Celts. The sisters were canonized by the Church and eventually many sacred sites throughout the land were named for them.

These sisters most likely are connected to the legend that survives around nearby Dundee that tells of the nine maidens of Pitempan. These maidens were devoured by a dragon or serpent at the sacred well now located in this parish. Many churches in the area are named for them, including a sacred well called Nine Maidens Well. Nearby, there is another sacred well called simply Nine Well, which is located on Finavon Hill.

Sacred wells and springs, especially those found near or in conjunction with temples, are places where rich troves of offerings were left, to be uncovered later by archeologists. These votive offerings include stone carvings, human heads, sacrificed animals, weapons, cauldrons, jewelry, effigies, coins, bowls, and other items left by pilgrims. In modern times clooties (strips of cloth or items of clothing) were used as votive offerings. Many traveled to these wells and springs from afar seeking healing or oracular work, to pay for a curse to be enacted, to give offerings to gods, or to bribe the spirits. Some came simply to pray. These sacred wells and springs were often seen as liminal places, as thresholds that provided entrance into the Otherworld where individuals could commune with the spirits who dwelled there. Sacred wells were considered to be places of healing, magic, and wisdom. They were said to be guarded by nymphs, faeries, and spirits, and were sometimes said to have spirits of their own.

Sometimes these water sources were dedicated to saints; sometimes they were dedicated to goddesses. They embodied powers that were both light and dark, both benevolent and baneful. On the one hand, they were used to grant wishes, to heal, and to foretell the future, as well as for divination and as portals to the Otherworld. On the other hand, they were also used to apply curses and to take life, and there are many accounts of ghosts or lost souls dwelling within them. Some spirits were benevolent and welcoming, while others acted more like guardians of the wells. We see this in numerous accounts of faery women and ladies in white appearing near them. In one legend, the Lady of the Fount controls the weather by throwing water from a sacred spring onto a slab of stone with a silver vessel. In other European Pagan and folk traditions, water was often thrown on the last cut of the harvest, or the last person to bring the harvest home, to insure a good rainfall for the following year's crops.

Other important groups of women from Celtic lore who associated with sacred water are the well maidens, the cup bearers, and the well guardians. These figures held golden vessels similar to the ones described above. They were well known to tend the sacred springs and offer food and nourishment to passing travelers. These sacred women, however, suffered at the hands of greedy men who violated their sacred space and stole their sacred vessels, silencing their voices and leaving the area around the wells to become barren wastelands. We can surmise that these well maidens may have acted as oracles. Another figure mentioned in similar texts is Mererid, meaning "pearl," who is described as both a well maiden and a cup bearer.

With the coming of Christianity, these wells and springs, which once belonged to nymphs, water spirits, and the old goddesses, were re-dedicated to saints. Yet the original spirits remained; they just adopted new names. But Christianity blurred the line between benevolence and evil by demonizing strong women and the groups in which they often associated. Many accounts of sirens, Dionysian women, faery women, and priestesses were corrupted by the rise of Christianity and the suppression of the divine feminine. Sacred wells, once tended by water priestesses under many guises, faded in the popular consciousness and were only preserved by blending with the new Christian practices and folk traditions. Sacred priestesses took on the guise of saints, wise women, and witches. One example of this is Mother Shipton, a 15th-century prophetess and village witch who was deeply connected with a petrifying and wishing well found just feet from her birthplace in a cave on a river. Another example is Brighid, a goddess of many sacred and healing springs who is honored in Ireland, Scotland, Wales, and Britain. We see her cult continue in the worship of St. Brigit, who lived in a monastery in Ireland with other nuns, a clear continuation of the old Pagan water-priestess and well-maiden traditions carried on under the new religion.

There are many surviving accounts of priestesses associated with the sea and even a few correspondences with the number nine. The ninth wave was believed to have healing properties, while going beyond the ninth wave was said to bring you to a liminal place. There are numerous accounts of sacred isles located in the ocean where priestesses lived. The nine priestesses of Sena lived on a small island in the Britannic sea. They worshiped a Gallic deity and were called the Gallisenae. These nine were believed to have the powers of sorcery. They could stir up ocean storms, cast magical charms, and shapeshift into the animal of their choice. They were said to heal the sick and act as oracles, but only to those who would cross the sea to consult them, most often sailors. We see similar connections with Morgan la Fae and her nine sisters from Avalon.

Exercise: Regaining Sovereignty with the Well Maidens

Sovereignty is power and authority. Too often, we hide our power or give it away. In some cases, we let ourselves be robbed of our power by trials, tribulations, and trauma. But in order to serve the waters the best you can, you must be a full and complete vessel. If you find yourself powerless, lost or lacking sovereignty, or in desperate need of healing, you can call upon the well maidens to help you find your own lost golden cup. For this exercise, you will need a gold cup or bowl that can hold drinkable water. You may have to create one by painting the outside gold. If you do create your own, you can use the process as a moving meditation and focus your intent into the vessel. You will also need white roses and spring water. Dress all in white and decorate your altar with white and gold.

When you are ready, sit or stand in front of your altar with your golden bowl in its center, but empty and upside down. Place your hands on the bowl and focus on that which you have lost. Contemplate the wrongness of the bowl being upside down and empty; consider how it is not functional in this state. Now shift your thoughts to change; think about regaining your strength and power. As you do this, turn the bowl right-side up.

Place your hands around one white rose and see yourself as this rose—this beautiful pure and whole being complete in its perfection and flaws, complete in beauty and thorny pain. Contemplate the most beautiful flower in the world, and how it contains sharp blood-drawing thorns. Then place this rose, a representation of your beauty and power, into the bowl. (You may need to pre-trim the stem so that it fits.) Pour the spring water gently over the rose and into the bowl (bonus if the water is from a sacred source). Visualize your body, your cup, filling with nourishing healing waters.

When you are done, close your eyes and call upon the well maidens, asking for their wisdom. Give them space to speak to you and do anything they advise in your ritual. Then open your eyes, remove the rose from the bowl, and set it aside. Add a handful of white rose petals to the water. Stir the water nine times sunwise, then anoint your brow, eyes, lips, hands, and feet with the water. Take the bowl in both hands and raise it over your head. As you begin to pour the water over your head, say:

Well maidens of the past,
I call on you.
Please help me with this task;
What has been lost
Is now found.
My golden cup I claim
In sacred water's name.
At this magic hour,
I regain my power;
This is my will,

So mote it be.

Forever sovereign

I will be.

Take the rose you set aside and dry it. Place the golden bowl in a place of honor on your altar or shrine and use it to work future rituals with the well maidens. When the rose is dry, take a marker and write "I am sovereign," or "I am powerful," or something similar on it. Then wrap it in gold or yellow cloth (silk is best), bind it with a golden thread, and set it on your altar or shrine. If you have a well-maiden altar, place it there.

The Avalonian Tradition

Avalon is and always has been a place of priestesses. It is a place of healing and a place where practitioners were trained in the magical arts and as healers. Many believe that modern-day Glastonbury, a town in Somerset in the west of England, is the location that was once Avalon. Many pilgrims travel there today to connect with the goddess, with their own spiritualty, or with the healing arts. While the original site of Avalon now lies buried in the mists of time, Glastonbury does seem to be a liminal place that has strong affinities to Avalon and may in fact be a portal to this ancient site. Others believe Avalon was in Cornwall; yet others believe it was the Isle of Lundy. No matter your belief, the spirit of Avalon is alive today and calling her priestesses home. And of course, you can access Avalon from anywhere through meditation, journey work, or spirit flight. Perhaps you visit there in your dreams or in past-life memories.

Avalon, whose name means "apple" or "place of apples," is a liminal place of mystery. It holds within its mysterious environs cycles of death and rebirth. Water plays a big part in the lore of the island, which was said to be ruled by the chief of nine sisters, Morgan la Fae. In fact, there are three sacred wells at modern-day Glastonbury:

- *The White Spring* is a calcium-rich spring that leaves white deposits and is dedicated to the goddess Brighid and the faery king Gwyn ap Nudd.

- *The Chalice Well*, which is rich in iron deposits, is home to the Lady of Avalon, who appears as a white lady or faery to those in need of healing or relief. This figure is variously identified as Morgan la Fae, St. Aeswitha, Melusine, or other faery queens.

- *The Black Well* is an obscure and little-known well that survives in the crypt of the Lady Chapel in Glastonbury Abbey. It is called St. Joseph's Well, although it is, no doubt, a remnant of goddess worship. The well may once have been dedicated to St. Aeswitha, a local woman with many faery attributes, whose depiction in the small chapel nearby gives clues to her origin. In the painting, she is shown holding a rose and a Celtic cross, suggesting that perhaps her powers of healing are considered to be *sub rosa*–hidden under the rose.

Glastonbury is widely believed to be a place where the mundane world and Avalon merge. Perhaps it is a portal, or perhaps it is where dimensions cross. No matter your belief, the fact remains that there are sacred wells there that nourish the landscape and encourage connection to water priestesses and their sacred role. Are these ancient priestesses beckoning us back to Avalon through a watery portal?

Exercise: *Magical Avalon Water*

Magical Avalon water is made with three main ingredients: white roses to represent the White Well, red roses to represent the Chalice Well, and apples sliced horizontally so the pentagram shows. You can also add apple blossoms, which have special energies when collected in the spring.

For this water, you will need:

- 1 cup white rose petals
- 1 cup red rose petals
- 3 apple slices
- 3 cups of water

Let the mixture stand for exactly nine hours. Then drain and bottle the water. Add a preservative if necessary. If you have water from a holy well or Full Moon water, these make great bases.

Exercise: Avalon Healing Elixir

This healing gem and rose elixir is useful in Avalonian and healing rites.

For this elixir, you will need:

- 9 red rose petals or 3 full red rose blossoms
- 9 white rose petals or 3 full white rose blossoms
- Pink quartz crystal
- White quartz crystal

Combine the roses to create a flower essence (see instructions for direct and indirect methods in chapter 4). Add the quartz crystals and let the mixture stand for nine hours, then drain and bottle the water. Add a little brandy as a preservative if necessary.

Exercise: Avalon Tea

This tea is useful for connecting with the energies of Avalon. It can also facilitate scrying and vision journeys to the sacred isle.

For this tea, you will need:

- 3 dried apple slices
- 9 red rose petals
- 9 white rose petals
- 6 hawthorn berries
- 9 rosehips
- 3 pinches of peppermint
- 3 pinches of nettle

Steep the herbs in a tea pot and serve at the desired temperature. Honey is a delicious addition to this brew. You can also add a pinch of mugwort, but please do not add this if you are pregnant or on medication.

Exercise: Avalon Rosacea Triformis Bath

Use this bath to connect with the energies of Avalon and the sacred plants that grow in its landscape.

For this bath, you will need:

- 3 apple slices (add apple blossoms if you have access to them)
- 1 pinch of hawthorn leaf
- 1 pinch of hawthorn flower
- 1 pinch of hawthorn berry
- 1 to 3 handfuls of fresh red rose petals
- 1 to 3 handfuls of fresh white rose petals
- 1 to 3 handfuls of fresh pink rose petals

Fill the tub to your desired temperature. Add Epsom salts if you wish. Cut the apple slices so that they are perfect circles that show the pentagram at the center. Add the dried hawthorn ingredients ground to your desired consistency. As you do this, contemplate the liminal aspects of hawthorn as a hedge and boundary plant. Add the pink rose petals, focusing on yourself and seeing them as an extension of yourself. Add the white rose petals, contemplating the White Spring and its mysteries. Finally, add the red rose petals, focusing on the Chalice Well and its mysteries. Then step into the bath and take a vision journey through the mist.

Through the Mist

Mist is the product of air and water that usually (but not always) appears in the liminal hours of dusk and dawn. In myth and folklore, it is often where spirits are found. It can also act as the hedge–the place between this world and the Otherworld. Because it appears in the liminal hours and in liminal places like the seashore or riverbanks, mist is seen as a portal into the Otherworld. We see this in the popular lore about Avalon as well, which portrays mist as the border between the mundane world and the holy isle.

Morgan la Fae, whose name is believed to be a variation of Morgen from Old Breton or Old Welsh meaning "seaborn," is perhaps the most famous Avalonian priestess. She was reported by Geoffrey of Monmouth in his *Vita Merlini* to be a faery woman who was the chief of nine sisters. Other texts describe her simply as a priestess who resided on the holy isle of Avalon. She was depicted as having wings, and as being a shapeshifter who had the gift of healing and knowledge of herbs and curative ointments. She was also known as a great necromancer who had the ability to commune with spirits and part the veil between this world and the spirit world. We see her character demonized hundreds of years later in the Arthurian romances, where she loses her role as benevolent priestess and takes on the role of evil enchantress.

Another Avalonian water priestess is the famous Lady of the Lake, who plays a far more significant role in Arthurian legend than she is usually given credit for in modern interpretations. Like so many other water priestesses we have discussed, she is part of both this world and of the watery realms. She is a water spirit, a demi-goddess, and a water fae. She is an Avalonian priestess and a water priestess. Her name varies from source to source—she is known variously as Nimue, Viviane, Elaine, Niniane, Nivian, Nineveh, or Evienne. Some believe these names all refer to the same woman, while others think they are the names for some of the nine sisters.

Exercise: Parting the Mist

"Parting the mist" is a phrase used to describe the process of crossing from this world to the astral realms. Like the river and the hedge, the mist is a liminal place. It also represents boundaries and is a boundary marker in itself. In addition, mist is often used as a metaphor for fog—the "fog" of our own minds and mental chatter. Parting this mist, pushing away this fog, thus prepares us for meditation and deep journey work. You can use this metaphor anytime you like—in ritual or in meditation, before or during a sacred bath, or standing on a quiet shore at dusk.

This ritual can help you part the mist and enter the spirit realms. After a time, you may find that the Avalon tea (described on pages 34 and 35) is not necessary, and that you want to access the Otherworld at the beach instead of in the bath. This is a natural transition and you can use this visualization as a doorway to open and close your access to the spirit realm.

Begin by creating the Avalon tea and sipping it for a few minutes to help settle your mind, align with the energies, and ground yourself. Then fill your tub with water; if you like, you can add rose petals and apple slices. Anoint your brow with holy water; if you have access to the Chalice Well or the White Well, use those waters. If not, create the Avalon water described on page 33 and use it to anoint your brow, hands, and feet.

When you are ready, step into your bath, relax, and take yourself into a meditative state. Visualize yourself dressed in white, with your hair down and your feet bare, standing on the shores of a vast and deep marsh. In the distance, you see a boat moving toward you. As it draws near, you see a man in the boat. He pulls the boat up to the shore and beckons you to get in. (His name is Barinthus and he is analogous to Charon, who ferries souls on the River Styx. He is connected to Manannan Mac Lir and, in some cases, they are believed to be the same.) You get in the boat and Barinthus maneuvers it to the middle of the marsh. When you feel ready, stand up, raise your hands in a "V" like that of the alchemical glyph for water and say:

Waters of Avalon, draw me near
To the place that I hold dear;
Clear of mind and pure of heart
These sacred mists I do part

As the mist begins to clear, Avalon appears before you, a beautiful green island covered with apples. The boat brings you to the shore. You disembark and climb a hill where you hear running water. Follow the sound of water to a white or red well. Spend time there in meditation and contemplation. You can stop here or journey on your own to your astral temple (see chapter 4) or to other realms.

When your time is done, retrace your steps and return to the boat. Barinthus will guide you back. When you are well into the marsh, stand up and raise your arms again in a "V," then bring them together and slide them down to your chest and sides while you say the incantation below. Let Barinthus take you back to shore, and be sure to pay him with a coin or other trinket.

Now the waters are hushed,

Still the mists

And calm the flood.

Silenced are the waters now—

Hush water, hush;

Hush mist, hush.

Back to the shore;

Back home once more.

Focus on your breath once more by counting nine full deep breaths. Then wiggle your toes and fingers, and roll your shoulders to come fully back into your body. Slowly open your eyes. Stay in the bath till you are ready, then drain it and dry off.

This vision journey is an excellent way to build your spirit work with the sacred isle. Once you become practiced at it, try it without the bath and eventually without the tea.

Exercise: Self-Love Bath with Morgan

It is important that you love and accept yourself and the sacred vessel that you are, and this ritual bath can help. This bath is going to be *all* about you, self-love, sensuality, and Morgan la Fae! Morgan is associated with Avalon, the Isle of Apples, so, for this bath, you will need a few nice apples—choose red or a different color if you wish. Cut them in slices that show the five-pointed star at the center.

Morgan means "seaborn," which brings to mind a vast ocean where the waters churn up white foam as the waves break. This foam will be represented by milk in your bath, and also calls to mind beauty baths, mother's milk, growth, nourishment, and life force. Hawthorn is associated with love and the faery realm, while the red rose petals, the rose water, and the rose quartz promote self-love. These associations all come together in a devotional ritual to Morgan and to yourself.

I recommend doing this bath on the Full Moon, but the timing is up to you. If you are working with deep healing and trauma, you may choose to do this in a darker Moon phase. Or you can try it at both times and work different aspects.

For this bath, you will need:

- Several fresh, washed red rose petals

- 3 apples sliced so that the pentagram shows

- 1 cup of milk (nondairy milk, almond milk, or coconut milk if you are vegan)

- $^1/_2$ cup rose water

- Hawthorn flowers

- Rose quartz

- 9 drops freshly collected sea water or 9 pinches sea salt

- Epsom salts

- A mirror

Make sure that all your ingredients are set and prepared before you start your bath. You can blend them all in a little dish so they are ready to pour into the tub. If you have a Morgan shrine, you can use some of the items as offerings. Make sure your mirror is somewhere where you can reach it once you are in the tub.

When you are ready, cleanse and set your space and the mood. You can add candles–I love having tealights all around, but red and heart-shaped candles are also appropriate for this bath. Invoke Morgan and begin to fill the tub with water. Add any extra ingredients you feel are needed, then place the apple slices in the water while contemplating the flesh, meat, and seed of the fruit. How do you see yourself there? Then hold the cup of milk and contemplate the nourishment of the mother and the ocean. When the tub is full, turn off the water. Hold up the mixture you have prepared and invoke Morgan's presence. Then sprinkle the mixture into the water, step in, and relax.

Take your time and enjoy the bath. Indulge yourself–perhaps with a glass of wine and some chocolate. When you are ready, begin to focus on you, your body, and the way you feel about your body. Think of the places that you despise on your body and hold the mirror to that spot. Look in the mirror and focus on accepting and loving that flaw. Consider why you are so hard on yourself about it, and recognize those feelings as the opposite of devotion and love to yourself and Morgan. Focus on what you perceive as the flaws in your body and make an effort to change your perception of them. The mirror may show you something different from what you have contemplated before.

When you are finished, find three to nine amazing things about yourself–things that you *love* about your body. Focus on these, nurture them, love them more, and try to balance the energy between your likes and dislikes. You can stay in this bath as long as you like.

This is hard work. You may have to take several of these baths before you feel as if you have really started to shift your perception of yourself and achieve greater balance. Take this bath whenever you are feeling down. Focus on shifting your negative perspective into a happier more accepting one. I have found that tears are often prompted by this bath. You can let them fall into the water, or catch them in a vial and use them in later work. This bath can be adapted to work with Aphrodite, Persephone, or other goddesses.

The Sacred Grail

Within Avalonian lore, we find the persistent symbol of the Holy Grail. What or who the grail is has been disputed for hundreds, perhaps thousands, of years. It has a few interesting connections to Glastonbury, where the Chalice Well gets its name from the belief that Joseph of Arimathea came to Glastonbury with the Holy Grail and placed it into the well shaft. But remember: this is not the only sacred well in Glastonbury. The White Well lies just a few feet away and the Black Well, or St. Joseph's Well, is located within the ruins of the crypt. The symbolic grail plays a prominent role in Arthurian myths and King Arthur's quest, and we see another version of a magical vessel in the story of the Lady of the Fount and her silver bowl, as well as in the stories about grail maidens and their magical golden vessels.

In other Celtic legends, we see similar magical vessels—for instance, in the story of Cerridwen's cauldron and in the accounts of the cauldron of rebirth kindled by the breath of the nine maidens. And there is another interesting connection with Glastonbury and magical or sacred vessels in the story of the Glastonbury blue bowl.

In 1906, a blue bowl was uncovered at Bride's mound in Glastonbury. Expectations among grail enthusiasts were high, but it was not the artifact they had hoped for. After carbon dating and other tests, researchers found that the bowl had been placed there in 1888 by an English doctor named Goodchild, who had spent time in Italy years before treating tuberculosis. He purchased the bowl there and brought it back to England with him. After a profound dream, in which he was told that Jesus had actually carried the bowl, he became convinced that he was in possession of the Holy Grail. He brought the bowl to an antiquities dealer, who could neither confirm nor deny its origin, but suggested that it was very ancient.

Since the person from whom Goodchild originally bought the bowl told him that it had come from inside a wall at an ancient site, he became even more convinced that it was, in fact, the Holy Grail. So he placed it in Bride's mound, where he thought it belonged. Then, in 1908, a Frenchman named Wellesley Tudor Pole felt compelled to move to Glastonbury. On his way there, he became convinced that he would find a sacred object there and would need three maidens to help him retrieve it.

We now know that the bowl was not the Holy Grail, and that most of the information put forth about it was incorrect. Yet the bowl still carries the energy of Bride's mound and the Chalice Well, where it now resides. Since I first read about this legendary bowl, it seemed right to keep something similar on my altar. The blue of the bowl connects me to the Third Eye, the waters, and the mysteries of the deep. The feminine shape of the bowl, and the fact that it is a womb-like vessel that holds water and life, connects me to the goddess. And so this vessel represents the goddess, healing, and second sight. For these reasons, it has become the central focus and main tool of my healing altar.

Lost Lands

No discussion of ancient priestesses would be complete without examining the energies of Atlantis, Lemuria, and Mu–and perhaps other continents, times, and cultures that have been lost to us as well. Even if these places are only hinted at in the historical record, more and more people are coming forth with memories of them from past lives. These manifest as a deep longing for these mysterious lands, which seem to have a place in our collective unconscious. I myself have experienced a haunting from the lost continent of Atlantis and a curious fascination with Lemuria, so I know that their energies are present in me.

Other practitioners, both men and women, have had vivid dreams of these lands and claimed profound past-life memories of them. I was skeptical of these at first–not so much of the existence of the places, as it seems natural to me that they do or did exist in some distant past, but rather of the sheer intensity of my impression that I had been there. I can't describe the longing I felt, the deep pull from somewhere within myself. No matter how hard I tried, I could not shake it off, nor could I explain it away.

An interesting connection between one of these lost worlds and Avalon, Glastonbury, and Morgan la Fae is found in the work of Dion Fortune (1890–1946), a latter-day British mystic whose famous novel, *The Sea Priestess*, has inspired many to take up service to the water as modern sea priestesses. The book's main character is, in fact, Vivian le fae Morgan, a reincarnated sea priestess from Atlantis with connections to water, lake ladies, and Avalonian priestesses. Fortune began experiencing visions of the lost continent of Atlantis at the young age of four. She identified these visions and later reported that they were deep memories from her past life.

Fortune spent part of her childhood in Weston-super-Mare, a seaside town located in the British county of Somerset. In her later years, she returned to Glastonbury–also in Somerset–and founded the Fraternity of Inner Light. Toward the end of her life, she went through a series of spiritual experiences in which she channeled "The Arthurian Formula," which she claimed derived from ancient memories passed down from the descendants of Atlantis when they arrived in Britain after the fall of that city. *The Sea Priestess* has influenced many modern practitioners and encouraged them to develop their own practices based on those of its heroine–a priestess of Isis and a modern re-incarnation of Morgan la Fae.

Many people believe that Mu is just another name for Atlantis; others believe it was actually Lemuria. But it may also have been its own lost land, as it is most widely believed to have been in the Pacific Ocean. Mu also shares interesting correlations with water. Its name comes originally from the ancient Egyptians, who used a version of it as their hieroglyphic symbol for water. Later, the name was simplified and used by the Phoenicians, who also translated it as "water." Mu is also the twelfth letter in the Greek alphabet.

Another interesting link to lost lands and past lives appears in Dorothy Louis Eady's account of Omm Sety. Eady was a fascinating woman who lived in the early 1900s who insisted that she was a reincarnated Egyptian priestess, Omm Sety. After falling down the stairs at the age of three, she began to express the conviction that she belonged to a different land. She was kicked out of school and Sunday school for her strange and vocal interest in Egypt. In her visions, she was visited by a spirit named Hor-Ra, who relayed the story of her first life in a text comprised of seventy pages of hieroglyphics. According to this text, she had been a priestess in ancient Egypt. After breaking her vows with her lover in this past life, she committed suicide to save him from scandal and a trial. Eventually, Eady traveled to Egypt, where she worked with Egyptologists, helping them refine their work and making accurate field observations as if she had seen it all before.

If Eady's account is true—and the more I look into it, the more convinced I am that it is—it seems to show that many of us may be capable of recalling past lives. In the case of Omm Sety, Eady was able to walk the land and accurately point out where the temples were and read the hieroglyphs. In the case of Fortune's Atlantis, however, the city cannot be explored and its writings have all been lost. So all we have are the memories, perhaps magnified and reflected back to us from the depths of the sea and only accessible to us through past-life recall and spirit journeys.

Yet there is no doubt of a watery connection. The presence of these lands ripples through our consciousness and reaches out to each of us, calling us back to the way of the water priestess, inviting us to heal the ancient energetic rifts, and challenging us to remember who we once were. Our task is to revive those old ways in our modern society.

These are controversial topics, and it is true there are many theories and conflicting thoughts and memories out there on the subject. But one thing that those who feel the pull of the energies of these ancient lands have in common is their conviction that they are being challenged to heal the planet and bring back a sacred awareness of the waters and the sacred arts. Many believe that they are working with guides and spirits who are linked in some way to these ancient times and places so that they can be healed, and then give that gift of healing back to the world.

You can connect with these ancient places and the priestesses of old by introducing a beautiful seafoam green color and Lemurian seed crystals to your altar. Bathe with Lemurian seed crystals and listen to the energy they put forth as you gently stroke them to connect to these realms and commune with their energies. Add beautiful blue stones like aquamarine and larimar from the Atlantic Ocean to connect with Atlantean energies.

The Vesica Pisces

The Vesica Pisces is both an ancient and modern symbol. It consists of two equal-size circles that overlap. But it is not to the two circles that we refer when we speak of the Vesica Pisces. It is rather the space where they conjoin that creates the sacred symbol. The name of the symbol means "bladder of the fish," which connects it to the ocean and other living bodies of water. When this symbol is repeated over and over, it creates the flower-of-life symbol. It was first mentioned in 300 BCE in the *Thirteen Books of the Elements*, but is widely used today in crystal grids and in the flower-of-life symbol.

This symbol represents the "womb of creation" and resembles the shape of a vagina. It appears often in medieval iconography and manuscripts and is found in many cathedrals and in carvings of the medieval Church. It is also connected with the Templars, with Freemasonry, with the Western Mystery Traditions, and with Glastonbury. The Vesica Pisces is also present in the Triquetra, or the trinity knot, where it is used three times to form the knot. For all these reasons and so many more yet to be discovered, the Chalice Well Trust decided to create a Vesica Pisces pool in the Chalice Well gardens.

Exercise: Connecting with Ancestral Priestesses

Use this ritual to connect with some of the ancestral priestesses from the distant past.

First, part the mist as described. See before you a beautiful island, full of trees that glisten in the sunlight with apples like ripe rubies hanging from them. You see many amazing and beautiful things. You follow a pathway that is green and lush and soft beneath your feet, to your astral temple where your sacred astral waters dwell (see chapter 4).

Approach the temple and use the sacred water you see to cleanse yourself. Then step into the sacred temple. State your intent to work as a water priestess and a guardian of the sacred waters. While you are in the temple, notice that there are other priestesses there as well; they are the water priestesses of the astral realms. Some are ancestors; others only tend the astral waters; others still look young and yet to be born into this realm. Seek the guidance of one or more of these priestesses and ask for their blessings.

Take any instructions you receive for the care of these waters. State your intention as if it were an oath. If you intend to visit once a month and clean the rivers or ocean, say so. If you intend to use the water to heal others and the planet, say so. Use words that are true and genuine, and only make promises that you can keep. Don't set yourself up for failure with grand ideas. Just be practical, and be sure to take on only what you know you can accomplish. If the priestesses accept your oath, they will give you a blue scarf or mantle. You may wish to purchase a blue cloak or robe to act as a physical representation of your devotion and to work with it in the physical world as well.

When you are finished, ask any final questions and then begin to journey back to the shore. Let Barinthus guide you back through the mist and be sure to close the mists the way you practiced above. Come back to your physical state and begin your work today. What you stated you would do, do. Start it now and be sure to come back often to continue your work. You can use this vision journey to travel through the mists to other locations in the astral realms. When you clear the mists, you may find yourself at a different location, or you may find other locations within your temple. When you are ready to close your journey, always return to the boat and come back the way you came.

Chapter 3

ANCIENT TEMPLE ARTS

ecoming a water priestess does not mean that you become a perfect vessel, have a perfect body, eat only organic vegetables, or spend copious amounts on special machines to create specific types of water. It means that you are constantly trying to be the best person and priestess you can be. Your body shape, a number on a scale, or your personal appearance have nothing to do with your work as a water priestess. Our work is so much more than that; it is far deeper than these superficial notions.

The way of a water priestess is that of the mother and requires the wisdom of the crone. It is built on a lifetime of learning and service to the water. Above all, the way of the water priestess is one of challenge, of service, and of commitment to ensure the safety and endurance of our waters. We can do this, not only by studying the ancient temple arts, but by looking at them critically and taking what we can from them to use in our own rituals. Your goal, however, must always be to create your own unique practices by reviving the craft and remembering the ancestral priestesses who once held water as sacred and lived their lives in attendance on it.

As you have seen, the arts of the water priestess are vast and varied. They include oracular work, shapeshifting, ritual bathing, goddess worship, the giving of offerings, temple tending, and divination. In this chapter, we will look at how these arts can be used in or adapted to a modern water-priestess practice.

Many things that modern-day water priestesses do as part of their work draw on these ancient practices. They may focus their energy on one particular thing, like singing or dancing in devotion to the water. Or they may layer up several different activities, depending on the intent and focus of their work. Here are some of the activities drawn from the ancient temple arts that modern-day priestesses can incorporate into their workings.

- *Body work:* Many priestess and healers take up massage, incorporating water to create a truly unique type of session. This may include ritual bathing, blessing with holy water, or calling upon the water spirits for help. You can practice Watsu or aquatic massage, or use hydro massage, or create sacred space for floating in a specialized tank. You can do this in your water temple, or in a sacred space you set up around it. In recent years, aquatic massage has become popular with many massage practitioners, who combine their skills with the power of the water.

- *Cleansing:* Cleansing is an essential part of any spiritual practice. When you live with one foot in the spiritual world and one in the mundane, you can attract positive or negative energies and attachments. If you are a healer, regular cleansings are especially important after you have worked with a client. You may also wish to practice regular cleansing rituals for those who seek your services. These can include energy work like Reiki, which uses water as a cleansing or enhancing element, or asperging or saining with sacred or holy water. Or you can create blessed herbal teas and bath salts that clients can use to perform these rites at home. You can even include some holy water with them.

- *Counseling:* A good portion of priestess work consists of listening to those who come to us for service or help. Even as card readers or energy workers, we often find ourselves offering counsel or listening to the traumas and trials of our clients' lives. If you find that you are drawn to this type of work or you find yourself in this position more often than not, consider becoming a certified counselor or an ordained minister, or seek out classes on how to hold space and support others.

- *Daily devotionals:* These are probably something all water priestesses should do in whatever manner is pleasing to them and aligns with their path. You will find water devotionals in the coming chapters that you can add to your daily practice.

- *Devotional dance, singing, and music:* Dance or music, when performed in devotion, is a wonderful way to honor your water practice and those who guide you along the way. Depending on your intent, you can sing for healing and drum furiously for protection; or you can dance for love near the water's edge. Your dance may focus on sacred movement or be a water prayer expressed through movement. Try singing your prayers or drumming during your incantations.

- *Energy work:* There are many forms of energy work; Reiki is currently very popular. Water, as we know, can hold, enhance, and transfer energy through its flow. These properties allow you to use water in your energy healing work. You can use water charged with Reiki energy to cleanse your subject or perhaps remove negative energy by releasing it into a bowl of water and disposing of it. Or you can work with the waters that our sacred body vessels hold by energetically restructuring the water already held within a client.

- *Goddess worship:* If your path focuses on worship of and reverence for a water goddess (or god), it probably incorporates many of the practices listed here, as well as some others you will read about in this book. The main focus of goddess or spirit worship is not only to honor a particular spirit, but also to spend time with the spirit in prayer, in devotion, in the crafting of objects, in the giving of votive offerings, and in the dedication of dance or song. You may also hold regular circles or rituals working with your goddess, in which you become the conduit between the spiritual waters and the people. A part of your practice may also be to mentor other devotees or priestesses, or even to create a permanent temple to this spirit.

- *Healing and wellness rituals:* You may find that you are drawn to the healing powers of water and want to offer water rituals that focus on well-being and health. You can perform these rituals in a river, a lake, or even a pool. Or you can create sacred space at a community center or in your living room in which to facilitate water healing circles that employ any number of exercises found in this book.

- *Meditation:* Meditation should be part of every magically and spiritually minded person's daily practice. However, you may want to take this deeper and offer meditation sessions in which water is used as a sacred element. You can record meditations with water creatures, spirits, or deities and offer them to others or to your community. Or you may want to offer rituals in which subjects drink sacred water before entering a meditative space.

- *Oracular work and divination:* Ancient temples were well known as places where people could commune with the gods through priests or priestesses who functioned as conduits for the spirits. These celebrants provided all kinds of oracular and divination services. As a water priestess, you can cast and read the bones, or scry with water, or read tea leaves, or even practice ritual possession.

- *Prayer:* Prayer is simply verbal communication with a divine being. As a water priestess, you can communicate with the water itself, with water creatures, with water spirits like Oceanids, naiads, and merfolk, or with the great water goddesses themselves. Many practitioners spend much of their lives or devotional work in prayer. If you are one of them, consider making a kneeling pillow or bench that is painted or embroidered with water colors and symbols and consecrating it as a magical tool.

- *Purification:* Like cleansing, purification can take on many different aspects. But while purification seems similar to cleansing, it is slightly different. Cleansing is the act of removing unwanted energy, while purification is the act of creating pure clean energy. This can take the form of purifying a space, or even a person. It may include the sprinkling of holy water on a subject, or full immersion into a sacred pool or bath.

- *Ritual bathing:* Bathing can be a beautiful and meaningful practice. Not only is it spiritually healthy, but it has mundane health benefits as well. Epsom salts are known to relax the body, as are certain herbs and even crystals (see chapter 8). Ritual bathing can involve more than just taking dedicated and structured baths, however. As a water priestess, you can facilitate ritual baths that introduce prepared bath salts, oils, or other physical items to create a sacred bath in a tub. Perhaps you can bathe in a retreat near a river or in a spring that has water safe for bathing or on the shore where the ocean serves as the sacred vessel in which subjects are cleansed and healed.

- *Ritual facilitation:* As a water priestess, you may be called to perform large or small community water rituals that can be either long and elaborate, or short and simple. These rituals may connect people to a water goddess or focus on healing a local water source. They may include ritual immersions or meditations with water, or any other type of water ceremony.

- *Tending a temple:* This art can take on a few different aspects. You can choose to tend your home temple, or you may choose to set up a tent at a community event or festival in order to offer spiritual water services, water blessings, and healing rites to others. If you have an office where your clients, students, or friends come to receive your services, you can work there, using one of the several practices given here, or you can create a practice of your own that focuses on holding space for you, your water guides and goddesses, or your clients.

- *Tending sacred water sites:* Like the priestesses and water maidens of old, you may choose to seek out a sacred water site and begin tending it by cleansing it or doing energy work or rituals there just as our ancestors did. Or you may want to work with a site that is not yet sacred. In this case, your work may focus on tending the physical environment to bring sacred energy back to the landscape and make it a sacred place for future generations. Find a beautiful grove where a spring or little brook flows, or a particular part of the beach to do your work. Even if it has never been known as a sacred place, you can certainly work to establish sacred energies there.'

- *Village healing work:* This type of work pulls from the energy of the wise women–those who, in another time, may have been called priestesses or witches. Wise women offered charm bags, healing ointments, and, in the case of water priestesses, spiritual washes, sacred baths, herbal teas, and healing or sacred waters. If you practice Reiki or another type of energy work, you can combine it with water to create a healing practice that calls on both your skills as a spiritual healer and your ability to cleanse and purify energy.

As you can see from this list, there are many ways a water priestess can specialize and many ways she can incorporate sacred water into sacred practices. Use the exercise below to help you sort out which activities and specialties resonate for you and which do are not. There is no lineage or tradition or temple that can dictate the particular skills you offer as a water priestess. This is your path; this is your choice. Choose, and let the water lead you.

Exercise: Defining Your Priestess Path

Spend time in contemplation or meditation, focusing on each of the practices above. Which ones call to you? Which ones reflect your strongest skills? Which ones turn you off? Which ones intimidate you? Make a checklist for each one of them, then rank them in order of importance to you. Begin by focusing on the top three–you can move on to the others when you feel ready to add more to your practice. If you feel drawn to practices for which you may need training or legal certification, be sure to go through the proper process, taking classes and getting certified before you begin to offer these services.

Offerings and Devotional Practices

Offerings and devotional practices are a central part of any priestess path. For water priestesses, they connects us, not only with the waters, but with the divine essence and spirit of water itself. Water beings like nymphs, naiads, and the goddesses found in mythology, folklore, and archeology were believed to be generally benevolent in nature. Some stories tell of mermaids who were stranded and gave gifts to those who helped them. Some tell how they bestowed the gift of second sight, or made gifts of cattle or magical items. Roman baths and Celtic archeological digs depict the water as a place of mystery, the home of spirits, and a portal to the Otherworld. The west, which is often associated with autumn and corresponds to the element Water, was thought to be the location of the Western Gate, the gateway into the ancestral realms.

Because of these traditions, the practice of placing votive offerings with the intention of never retrieving them was quite common. These usually took the form of physical objects like coins, effigies, pins, buttons, and jewelry. These items often turn up in archeological digs today. When you give offerings to the water, however, be careful how you select them as well as how you adapt them to a modern practice, as they will stay on your altar or be permanently moved to your water shrine.

Offerings that are deposited near or in the water are called votive offerings. Ancient people placed items like effigies, coins, buttons, pins, jewelry, weapons, and other items of importance in or near sacred places as offerings to their deities or spirits. Sometimes these were deposited into lakes; sometimes they were buried; sometimes they were left in or near a sacred temple. Today, this is not either practical or safe for our environment, as our waters are already polluted. Adding objects and possible contaminants to them, even if they are given with good intentions, isn't helpful and is completely counter to the work we are trying to do. If there are occasions on which your ritual may call for this type of offering, I urge you to think deeply about what you choose to deposit in the water. While flowers may seem harmless, store-bought flowers often contain pesticides, dyes, and other chemicals. Consider growing organic flowers if you need to use them in your rituals. This will add additional intent to the offerings and you can even water the plants with Moon water or other sacred water.

Drawing sigils in the sand and creating sacred geometry with river rocks or shells is another way to leave offerings without negatively impacting the water or the environment. You may even find that the river, lake, or ocean where you work or visit contains stones like quartz, jasper, or agate. If you have an ethically mined stone that hasn't been treated or dyed, you can give this as an offering imbued with healing or protective energy. Consider giving only tumbled stones, which are rounded and smooth, as points and sharp edges may hurt others who walk your shores and waterways barefoot. Try to use only stones that are naturally found in the area and give your offerings the added gift of energy, prayers, spell work, or sacred symbols.

A third type of offering consists of actions or deeds, rather than physical forms or objects. These offerings can be given safely, and should perhaps become a focal point for your devotional work. They include energy work, cleaning up riverbanks or ocean shores, protection magic for water creatures, joining protests, cleaning up oil spills, and any other actions that show your devotion and gratitude rather than polluting the water even more by adding foreign objects. You can sing, draw protection symbols in the sand, or create a devotional dance to provide healing energy to the water.

Water Spirit Guides

No matter what type of water priestess you consider yourself to be, you will most certainly have guides in this realm and in the Otherworld. In the mundane world, these guides may take the form of a parent, an elder, or an educator. They may also appear as omens in nature–strange animal sightings or natural phenomena that go beyond just coincidence. As a priestess with the ability to work with the spiritual and supernatural worlds, however–and often acting as a conduit for the water or a sacred cause–you may find yourself in need of advice or guidance from a source you can't readily see. This is when your spirit guides can show you the way, gifting you with supernatural powers or aiding you in healing work, divination, or ritual. They can provide wisdom and guidance, as well as spiritual support.

The first thing to be aware of when seeking your spirit guides is that not all spirits are benign and not all are authentic. Spirits can and will act just as humans do. Always proceed with caution and ask the spirits when you first make contact if they have your highest good at heart. If they don't, move on and find a spirit who is beneficial to you. The second thing to keep in mind is that this is a *reciprocal* relationship and not a one-sided association where you just ask for help but never give back. So before you are given any gifts–or better yet, before you even ask for them–it is better to establish a relationship of reciprocity.

There are many ways spirit guides may manifest. In fact, you may already have one. If you work with angels, faery folk, land spirits, ancestors, priestesses, mermaids, dolphins, goddesses, or even with water creatures like the intimidating megalodon, they possess ancient wisdom and primal power that they are willing to share with those who are pure of heart and want to work in service to the water and the spirits that dwell there. You may also have contact with water spirits from your ancestral lands or a particular body of water. You may speak directly with the energy of a waterfall or the lake fae, or with a well spirit or any other form of water. Any of these and perhaps some I have not mentioned may act as your spirit guides.

Like water, these guides come in all shapes and sizes and may be permanent or flow in and out of your life. Moreover, they are known by many names–angel guides, spirit guides, familiars, spirit lovers, and many others. It is important to keep an open mind. More often than not, we don't get to choose our guides; they choose us! They may take a form that is pleasing to you, or they may show up as something scary. And they may appear in the form you least expect.

Moreover, not all encounters with spirits are going to be positive, and not all scary or negative encounters with a spirit mean that the spirit is evil. If you are apprehensive about contacting spirits, you can always seek help at your local metaphysical or Pagan shop, or even read books on how to work with or deal with spirits and deities. If you encounter an aggressive spirit in a natural place, consider whether you may be the one in the wrong. You may be invading a sacred grove or a place untouched by humans in which spirits dwell. It is possible that the aggressive energy you are sensing is simply protection against an outside invader–you. In some cases, you may encounter a spirit who is full of light, beauty, and bane. In other cases, a spirit who seems scary may actually have a benevolent nature. Always protect yourself, and always go cautiously into these workings. This will also help you to introduce yourself to a spirit in a humble way that will nurture and not force the relationship. However your guides reveal themselves to you, be sure to approach them with honor and reverence. If they give you their names, you must keep that secret, as names have power. To be given the true name of a spirit is a sign of trust.

Exercise: Finding Your Water Spirit Guide

You may already know you have guides, or you may already have been guided by them. Even if you have guides, however, this exercise can be helpful. If you don't have guides, use it whenever you need it for specific tasks or rituals, or to gain insight from different types of water, water goddesses, or ancestral priestesses.

You can try this by gazing into a lake or pond, or on your altar by scrying (see chapter 7). Remember that there is always the risk of astral nasties showing up when you call on the spirits, so always be sure to create a protective bubble or circle wherever you are working (see chapter 4). After you have protected your space, you can begin.

As for all meditations or spirit journeys, make yourself comfortable–but not too comfortable, as you don't want to fall asleep. Read through this spirit journey several times so that you are familiar with it and can do most of it from memory. You want to be in a meditative state, and reading can interfere with that.

Begin by breathing deeply and fully for nine breaths. With each breath, focus on an ancestral priestess and her attributes–not on any particular one, rather on an abstract version of the priestess. Allow your mind to wander and pick up on anything that comes to it. If this is too much for you, just begin by focusing on nine individual breaths, knowing that the number corresponds with priestess groups, which often appear in clusters of nine.

After your ninth breath, let your mind become blank and visualize only darkness–the darkness of the subaqueous realms, the depths of the ocean, and the darkness of night. Allow the darkness of night to be illuminated by a thin sliver of crescent of Moon and the twinkling of stars. Then lower your gaze slowly and let it melt down toward the earth. When you reach the horizon line, keep going until you see the night sky reflected in a glassy lake surrounded by sacred trees and standing stones. Pay attention to the standing stones and any symbols that may be carved into the rock. Take notice of the trees and observe them. Is there anything special about them?

Allow your gaze to sink down and begin to focus on the reflection in the water. Notice the stars and anything else that may be illuminated there. Speak to the lake and ask it to reveal a spirit guide for you, one that belongs to the water and is willing to guide you in the way of the water priestess. Be sure to state firmly that only spirits of good will are to appear. Gaze below the surface of the lake and see if there is anyone or anything swimming below.

Ask the spirits if they are willing to guide you on your path. If they agree, ask how to contact them in the future. You can also ask about any specific signs or means of communication they may use so you can be sure you are connecting with the correct spirit. You can ask their names, but you may find they are reluctant to give them. They may give you a nickname, or they may ask you never to share their names. This is something you must always respect.

Allow yourself to drift out of the meditation slowly, leaving in the same way you came. Begin by focusing on the reflection of the stars and Moon on the surface of the water, follow the light back to the horizon line, and keep going till you are gazing up at the night sky. Allow the stars to dim one by one and the sliver of Moon to fade, until you are again gazing at dark nothingness. Focus back on your breath for nine counts, then slowly begin to move and come back into this time and space.

If no one appears, thank the water for its connection and try the exercise another time. It is possible that you are not ready for your guide at this time. Perhaps you need to cultivate a better relationship with water through regular offerings of time, prayers, healing work, or cleansing. These can deepen your relationship and encourage the water to connect with you.

Exercise: Water Creature Meditation

This meditation is best done on a Dark Moon. In it, you will travel via meditation to the bottom of the ocean, lake, or river to see which water creatures are willing to align with you for your work, or possibly for your lifetime. You may see the creature in the water or in your mind's eye. Once you complete the meditation, do some in-depth research into the creature revealed. You must get to know this animal before working with it. You must know who you are working with, and why. There are so many possibilities and each person is different. Part of your power as a water priestess lies in uncovering your own personal egregore and the mysteries that come with it.

During a Dark Moon, choose a small dark-colored vessel to use as you communicate with the water spirits. Set your altar or work space with nocturnal energies—black and dark-blue candles, deep-toned earthy music, and perhaps even black salts, flowers, or stones. If you are journeying to the ocean, add sea water and some seaweed; if journeying to a lake or river, add fresh water and perhaps one plant that grows locally. You will sit or stand with this bowl so that you can gaze into the water.

Begin by anointing your Third Eye with the water, and perhaps your palms and heart center as well. When you are ready to slip into meditation, close your eyes and begin to breathe in a way that will help calm and relax you. Spend time just listening to your breathing; listen to the inhale, then the exhale. Soon these breaths will begin to sound like waves lapping at the shore.

In your mind's eye, all is dark and you are sitting on the banks of your favorite sacred water source. You are comfortable and alone. You can hear the water moving, gently lapping at the shore, but you see that the surface is calm. Take this time to speak to the water. Ask it if you have a guide or ally for this journey, and ask that it be revealed in the surface of the water. You may see a reflection, or you may see the creature(s) emerge from the water. Spend time with your creature(s). Ask how they can help you in your work and if there are any special skills they can lend or even help you to develop on your own.

When you are finished with your interaction, thank the creature(s) and give an offering of some kind. Let them return or fade away, then sit on the banks or the shore once more, letting the water grow calm, until it has a smooth surface once again. When you are ready, focus on your breathing and return back to your own space and time. Record your meditation in your Book of Tides.

Exercise: Communing with Water Spirits

Once your temple has been set up and cleansed (see chapter 4), you can use this ritual to commune with your water creatures, or with your spirits and goddesses. Here, the ritual is specifically written to contact and work with water spirits, but you may want to perform it again after your initial try while focusing on a specific type of spirit. Do you want to speak with ancestral water priestesses? Mermaids? Lake ladies? Benevolent water horses? Oceanids? If so, set the imagery on the altar to align with the type of spirit you want to attract.

Begin by protecting your space. You can cast an astral circle or bubble, or set up protective talismans, or use salt, sand, or stone to protect your working. Be sure to take the extra time to set up your protection for this ritual, because you will be seeking out spirits in the Otherworld and not all of them may have your well-being at heart. Although you will request that only benign spirits appear, it is always good to have a back-up plan.

For this ritual, you will need:

- A small circular mirror that can be submerged in water

- Altar

- Altar cloth

- Carefully selected water

- Journal and pen

- Salt for your circle

- White and blue candles

- A sacred vessel, preferably a bowl (a darker color is better, but not necessary)

Take a few deep breaths and bring yourself into a light ritual trance or meditative state. You may want to sway like the water, or just clear your mind and float into the Otherworld. Be sure to seal your protective barrier in the astral realm. Then place a few drops of your water on the index finger of your dominant hand and draw the alchemical symbol for water (a triangle with point down) onto the mirror. Hold the mirror to your mouth and blow on it to fog it up. Then draw your own sacred symbol (see chapter 9) on the mirror and bring it to your mouth so you can fog it up again. Draw a sunwise spiral that goes from the center of the mirror outward.

Say the incantation on page 74, then place the mirror in the water facing up. Sit or find a comfortable position and gaze gently at the mirror. Let yourself drift into mediation and begin to see images in the water or on the mirror. Your eyes may slowly shut and you may be whisked into the Otherworld to commune with the water spirits. When this happens, the water portal is open and you can commune directly with the great water goddesses or water spirits that come through. This portion of the ritual will flow uniquely to and from you. Once you feel you are finished, come back out of your altered state and record what you saw in your Book of Tides so you can continue to develop your relationships in your devotional and ritual practices.

End the ritual by closing the mirror. To do this, place the index finger of your dominant hand into the water on top of the mirror. Starting at the outer edge and moving in a counter-sunwise direction, spiral your finger toward the center of the mirror. Then seal the mirror by drawing your sacred symbol and the alchemical symbol for water. Close your circle or sacred space and remove the mirror. Wrap it in a black cloth and store it to use another time. Immediately take your water outside and offer it to the earth—perhaps a sacred rose bush or other watery flower you grow.

Water Spirit Incantation

Spirits of the water, I draw you near
To this sacred space I hold dear.
Water spirits, I call you here
To this watery portal with no fear.
Water spirits who are of bane
Are banished from this space never to appear again.
Spirits from the water's sacred place,
Come forth to show your holy face.
I bridge the realms through portal divine
To commune with the spirits I call mine.
Sacred waters, show me a sign!

Protecting the Waters

Part of tending the waters is caring for the spirits who dwell there. This includes astral beings, as well as animals that exist in the mundane world. Each day, we see more and more evidence of the abuse the water is taking from a greedy human race. Sometimes it is an oil spill, sometimes a new pipeline being forced upon the land, sometimes a dolphin slaughter that takes place. We feel hopeless when we encounter these things and wish we could help, but are often left wondering how. Not all of us have the financial means or the time to jump in our cars and dedicate the next two months to a clean-up or a protest. So what can we do?

There are actually a few options. First and foremost, we can pray; we can pray to the water spirits, to the great goddess, or to Gaia herself. We can pray for the planet and for those who are harmed. We can even pray that the minds and hearts of those who are so destructive can be opened and changed. We can send healing energy to these trouble spots and, finally, we can focus on protection magic.

There are many ways to perform protection magic—with tools, in the astral plane, through dreamtime journeys, and through ceremonial rituals. The following exercise is a small ceremony you can perform as a protection rite. You may perform this as part of your already established rituals, or it can be performed alone.

Exercise: Water Protection Ritual

If you want to protect specific water creatures like dolphins or whales, start by printing out an image of the creature or location you want to protect. You can also use an effigy or a drawing of the animal. Then place a clear glass bowl on top of the image on your water altar. In the case of an effigy, place it in a smaller bowl and place that bowl in the larger glass bowl so the effigy doesn't get wet. Then add water to the larger bowl and pray over it or send energy to the water in the bowl.

Do this for nine consecutive days. On the ninth day, take the water to the nearest water source and pour it out. If you are working protection for ocean creatures, the protection water must either go into the ocean or into a river that runs to the sea. If you are working for a fresh-water lake, river, or spring, deposit the water as close to the source as you can. Make sure you are working with clean water, so you don't pollute the source with chemicals or processed water. You can also use water directly from a location; just be sure to return it there so you neither harm the area nor permanently remove the water. And be sure to let the water spirits know why you are taking the water and that it will be returned. At first, you may want to focus on a single body of water and the creatures that dwell there.

There are many ways for you to work protection magic for a particular local water site. I recommend visiting the site yourself to perform the ceremony, but if you can't, consider using a photo, dirt, sand, or water from the site, or a map or GPS location in your own water temple. If you are working for a lake, walk the perimeter and cast an energetic circle of protection. If you are working for the ocean, draw protective symbols in the sand or create them with shells and stones. There are many ways to incorporate protection for our waters into our practice. All we need is love—and intent.

Chapter 4

RAISING THE TEMPLE

A temple space is central to a water priestess' practice. But it is important to note that a water temple can be anywhere–in a dedicated room, at the seashore or by a river, in a corner of your bedroom or in your bathtub, at your kitchen table, or in a dedicated space outside in your yard. It may even exist in your mind's eye.

To find the best location for your temple space, spend time in mindful meditation and ask the water spirits where it should be. Once you select a location, perform a quick cleansing ritual, like the one given on the next page, to prepare the space. After your space is completely set up, I advise doing the longer version of the ritual using your newly consecrated sacred vessel. To quickly clear the energy, you can either asperge with holy water, using a sprig of herbs to sprinkle the water all around the space, or you can burn an incense of myrrh, a resin associated with the element of Water.

To set up the temple, you must create a sacred space in which to perform your devotional work or facilitate healing rites or energy work for clients. I have found that keeping things like blankets, pillows, other bowls, jars of water, and beautiful adornments in the temple space can be useful for making yourself comfortable in your work, or if you are working with a client or student. Of course, how you decorate it is up to you!

Sometimes you may need to set up temporary temple space, for instance if you are performing a ritual near a water source, at your local metaphysical shop, or even in a public place. Begin by cleansing with the four elements in conjunction with water. First, circumambulate the space with the element of Air, perhaps using a feather you found at the beach with the smoke from myrrh incense. Then circle again with the element of Fire, perhaps using a blue candle anointed in holy water. Circle a third time with the element of Water, perhaps sprinkling it all over the space as you circle. Make one last pass with the element of Earth, using either sea salt or a little sand from the beach. The order of the elements can also be up to you. In my own tradition, we start in the north, the land of the ancestors and the dead, and we end with Water. In another tradition in which I was trained, we started in the east. In my own water practice, I often start in the west, which is the gate to the Celtic Otherworld. You can also place river rocks at the four cardinal points as you circumambulate the space. Once the space is cleared, cast a protective circle (see the exercise starting on page 85). Now you are ready to invite others into your circle or begin a solitary ritual.

Shrines and Altars

A shrine is a space set up to honor or venerate something or someone. In the case of a water shrine, it is a place where a water priestess honors water goddesses or ancestral priestesses, water spirits and nymphs, mermaids and other finfolk, and ancestral and live water animals. These spaces are generally stationary and regularly tended and cleansed. They are places where regular offerings are left.

Your shrine is a place where you perform your devotional practices. It is a space where you can place ritual items to be charged and blessed by the spirits that reside there. It can also act as a spirit house or as a place where a spirit house sits. Quite often, these spaces are adorned with shells, flowers, precious gemstones, images, and statues. (It is okay to use store-bought foods and flowers on your home shrines and altars.)

A water shrine is a space where you can worship or honor a water goddess or spirit, or just water in general. You can place your sacred waters there, and perform devotional offerings and prayers. You can also add shells or crystal grids to perform healing work. Feel free to design your shrine however you like and use it in any manner that suits you. The important thing is that it be functional and represent you. The only requirement is that it honors your spirit and is dedicated to water in some way. After all, this is your sacred space—a reflection of you that holds your energy and devotion to the water. It is where you will begin to focus your prayers and rituals.

Altars and shrines are often confused with one another, but they are actually different. Unlike shrines, altars are places where magic is worked. They may be taken down and reconstructed in different places–like at the beach, on the banks of a river, or at special celebrations and events–or they may remain in one place in readiness for a priestess to perform workings dedicated to a particular water goddess or healing spirit. I personally have altar spaces that are permanent and that sit ready for my workings. If you perform rituals for community circles, your altar must be set up in a particular manner facing a particular direction, and must contain specific items associated with the tradition you are following and the collective group mind you are accessing.

Many magical practitioners combine altars and shrines into one space, making the back portion more of a shrine space and the front portion more of an altar space. Since a water priestess's work focuses mainly on devotional practice, healing, and prayer rather than on magical ritual, I will focus here on creating a shrine space, but I have included an brief exercise for setting up an altar as well.

Exercise: Setting Up a Water Shrine

A water priestess needs a dedicated space within which to do her work. As your practice grows, you may find that you need both an outdoor and an indoor shrine, or you may find that you do most of your work at the water's edge. To begin, we will work on building an indoor shrine space, and then clearing or consecrating the temple area in which it sits. This may be a room or a corner in your home, or any other area you choose. The instructions will help you create this space and you may also find them useful for clearing and consecrating spaces used for public rituals and temporary circles.

First, choose a surface on which you can begin your work. This can be a table or shelf. If you choose to burn candles on your shrine, be sure to choose a space with an open top. A bookshelf may not be a good option here, but small end tables or night stands work really well. Once you have chosen your space, consecrate the surface the same way you consecrated your sacred vessel. When you are finished, place the table or shelf in the west and facing west. In Celtic traditions, the Western Gate is the portal where journeys into the Otherworld begin. West is also associated with the element of Water. If you cannot place or face the surface of your shrine in the west, try to orient it to the nearest body of water or the one with which you work the most.

You can place an altar cloth or pretty piece of fabric representing your practice on the surface. If you do this, consecrate it the same way you consecrated the shrine surface and sacred vessel. In fact, you may want to repeat this process for each of the items you use on your shrine and on future working altars.

Once that is done, choose a statue, image, or item that represents your practice or your cause and place it on the shrine. This will be the focal point of the shrine and will represent the deities or powers to whom it is dedicated. You may wish to decorate the shrine with candles, offering plates or bowls, flowers, gems, shells, stones, or anything else that symbolizes your intent. Just make sure you leave space for your sacred vessel, as you will be adding that next. Over time, you may find that things shift—for instance, you may decorate your shrine differently for seasonal celebrations or in conjunction with the phases of the Moon.

Exercise: Setting Up a Water Altar

Altars can be set up anywhere and can include simple or elaborate items and layouts. Items you can use on your altar include altar cloths, statues, candles, ritual tools, shells, and, for water priestesses, most likely a chalice or bowl. Most altars I have observed are more like working spaces and thus are often quite messy–but not dirty–whereas shrines are usually kept clean and lovely in devotion to the water spirit. Altars can take any form you like. The important thing is that you have a particular place to focus your work and your devotional practices. A clever water priestess may also want to create a water altar for other purposes as well– perhaps for protection or for another specific cause.

Blue, as the color most often associated with healing and water, should be the primary color for your water altar. To establish blue as the predominant color, choose a blue chalice, bowl, or other sacred vessel. When selecting healing items like crystals, herbs, symbols, or talismans to adorn your altar, choose blue when possible. Consider using the symbol of the Vesica Pisces found on the Chalice Well, as well as sacred healing waters stored in blue vials or jars, blue flowers and other offerings, and statues of the water goddesses you are petitioning for healing. It is up to you how your altar looks and what you put on it. Just make sure you leave space for your sacred vessel!

Exercise: Creating an Astral Water Temple

While you are in meditation in your physical water temple, you may want to have a place where you can commune with ancestral priestesses, faery women, goddesses, and other guides. This exercise will help you create a water temple within the astral realms to use for your magical and spiritual work. To do this, first visualize what it may look like, but don't try to plan it out in detail. No doubt your creative leanings will influence the form it takes, but so will your subconscious, your guides, and the water spirits who are working alongside you.

While I encourage you not to pre-plan your temple, you may already have begun visioning it, or you may even have visited there before. There are unlimited styles and forms these temples can take. You may find yourself in a Grecian temple, or by a sacred spring in a magical grove, or in a crystal cave or an Atlantian temple. You may find yourself in a sea temple like the one described in Dion Fortune's novel or perhaps in a place that is completely different in form.

Enter into a meditative state and allow your vision journey to take you to an ancient water temple. Observe what it looks like, where it is, and what activities are happening there. Once you have a clear picture of this in your own mind, visit it often so it begins to solidify within the astral realm. The saying "As above, so below; as within so without" applies here. You should enact the water rituals you perform in the physical world in your astral temple as well. Create your sacred waters and bless the astral temple using the methods shown throughout this book.

There are a few important things to consider when constructing and working in your astral water temple. As soon as it is constructed, go through a ritual to bless and protect it similar to the one you performed for your physical temple. It is important to protect yourself in both the physical and spiritual realms. When you are constructing this temple, it may appear instantly, or it may show up in pieces, or you may see it constructed right before your eyes. Time may move differently in your astral temple, and, if you end up with astral temple guides, they may be working while you are away in the mundane realm. Be sure to cleanse your astral temple space just as you did your physical temple space. And you can place astral offerings on its altar as well.

Exercise: Cleansing Your Shrine or Altar

You can use this ritual to cleanse either your shrine or your altar, as well as any other sacred space. In it, you will use a mixture of rosemary, St. John's wort, and white rose petals to charge sacred water, and then asperge the area you want to cleanse with an herb bundle dipped in that water. Dried and crushed rosemary and St. John's wort work best to create the cleansing water, but freshly harvested herbs and essential oils can work as a substitute. You can also substitute any pure white flower for the white roses. If you have access to fresh herbs, use them in your herbal bundle. If you do not garden, most grocery stores sell fresh sprigs of herbs that can be bundled, and leaves from a tree or tall grass bound together can work as well. If you don't have access to water from a sacred source, you can create your own, or use Moon water or spring water. You will also need a sacred vessel and a piece of white string to tie your bundle.

The night before the ritual, gather your choice of water and place it in your sacred vessel. Add the rosemary, the St. John's wort, and the white rose petals to the water, and let the mixture sit overnight. If your space is already prepared, you can place it on your altar or on your water shrine. In the morning, after your daily devotional practice, remove the vessel and strain out the plant material. Then put the strained water back in the vessel.

To begin the cleansing ritual, gather your fresh herbs and bundle them together using the white string. Dip your fingers into the sacred water and a tie knot in the string, saying:

With this knot, may the water bless this bundle.

Dip your fingers again and tie a second knot, saying:

With this knot, may the water spirits bless my work.

Dip your fingers a third time and tie a third knot, saying:

The power of water in trinity; may this work be blessed.

Set the bundle aside.

You are now ready to cleanse the energy of your altar or shrine and prepare it for ritual. Start by facing west, the direction aligned with water, then walk in a sunwise direction while chanting and saining. Dip the herb bundle into the water and gently flick the water through your space. Be sure to circle around the space nine full times, coming back to the west. While you do this, chant this blessing:

Spirit of water, come alive in this place.
Spirit of water, cleanse this temple space.
Spirt of water, bless this space.

You can use this ritual as often as you like and for many different purposes—as a dedication ritual, before baths, as a monthly ritual on the Dark Moon, before a large working, or even to create a sacred space for a group circle.

Protecting the Temple

When you follow the way of the water priestess, you walk in both the mundane and the spirit worlds. But when you shine your light in the spirit world, many types of spirits take notice—and they may not all have the best intentions. Even if you never acknowledge their existence, it is best to protect yourself and your shrine from these malign spirits. There are many old folk beliefs about forces like the evil eye or malevolent stare sending bad energy against someone. So if you are going to work in public, have a public presence, or participate in any type of activism or even healing where you work with other people's pain and trauma, you must protect yourself. You are no good to anyone if you are bogged down by other people's negative energy. Below are a few different types of protection you can use for both your person and your temple space, or perhaps in a healing circle with other people.

Exercise: Protection Bubble

A protection bubble is a kind of force field that you create around yourself to ward off malign spirits and negative energies. It can start in your center and move outward if you need to extend it to a circle or temple space. If you are working with a large circle or for a long period of time during which you may need to use your energy for ritual practices or healing, you can anchor your bubble with something like a ring of sea salt, quartz points, or bowls of water set in the cardinal directions. Expand your bubble out as far as you need to ensure that you and others are safe.

Exercise: Protection with Sea Salt

Salt has always been used for protection and many magical traditions use it in various ways to clear negative (baneful) energy, as well as to create protective barriers in the form of a circle. You can probably find fine-grain sea salt at your local grocery store, or you may want to use salt that comes from a particular body of water—for instance, Okinawan salt, Celtic salt, or Dead Sea salt. The important thing is that the salt flows easily. For instance, Dead Sea salt doesn't always flow easily and so does not work particularly well as a protective barrier. But it does work well for a protective bath. In fact, a little Dead Sea salt with rue, rosemary, and hawthorn flowers makes a wonderful protection bath, although women who are pregnant or actively seeking to become pregnant should avoid adding rue and rosemary directly to the bath water. Instead, create an indirect essence with the herbs and add the essence to the water. You can find directions for making direct and indirect essences below Rue can cause photosensitivity; use the indirect method, if necessary.

To create a protective barrier with salt, you will need a thumbtack (or a friend), a string half the desired diameter of the circle you want to create, and a container of sea salt. If the container the salt comes in is not suitable for pouring, transfer it to an empty condiment bottle–the kind found in restaurants that you can refill. These usually have pointy tips that direct the salt perfectly. I have also had success with children's sippy cups; just be sure the salt pours from the center. If you have to make the hole larger, use a Phillips head screwdriver.

Once you have the bottle of salt ready, tie the string around the bottle and make a small loop at the other end. If you are working alone, push a long thumbtack into the floor. For a water circle, start in the west, as this is the direction that is most often associated with water in esoteric traditions. As we discussed, however, you really can choose to start in any direction; just be sure you do so with purpose and intent.

Using the string as a guide, create your circle by letting the salt flow onto the floor. (You can do this on carpet, but it is best on a smooth surface and easier to clean up.) When manifesting, healing, working with solar energies, or performing positive magic, travel sunwise around the circle. For nocturnal work, underworld work, or journeying to ancestral realms, travel counter-sunwise. Counter-sunwise movement spirals down, while sunwise movement spirals up. Once the circle is complete, take a vessel filled with holy water (see page 88) and walk around the inside of the circle in the same direction, saying:

I consecrate this circle with the power of holy water;
I bless this circle with the power of the sea;
Waters, cleanse this space and protect me!

Repeat this nine times as you circle and end back in the west (or in your starting direction).

There may be occasions when you don't want to take all this time to create a perfect protective circle. In that case, just pour the salt to create a circle without using the string. Just be sure you come back to your starting point to close it. My very first salt circle looked more like a wobbly egg–and I used the string! Be patient with yourself and realize that this is a skill that can only be learned through regular practice.

You can also add herbs to the salt or place them around the outside of the circle. You can draw triangles (with the point down) to symbolize the alchemical sigil for water; and you can add personal protection sigils, shells, or crystals to the circle. These all enhance the working and increase the vibration, lending extra power to your water magic.

Exercise: Protection with Holy Water

The use of holy water dates back to pre-history. In fact, water was always held as sacred. We see this in the cults of the sacred springs and holy wells, as well as in Christian churches and other religious temple spaces. For many, water is part of spiritual practice. Muslin worshipers cleanse before entering the mosque; Christians baptize their children with holy water and Catholic priests use it in their Sunday rituals. And, for obvious reasons, holy water is essential to the path of the water priestess.

You can create your own holy water in several ways. If you have access to a sacred spring, a holy well, or even a church where holy water is kept in a basin, you may be able to use this. Of course, if you are unsure, ask permission first! Many sacred water sites allow the gathering of these holy waters. If the well is dedicated to a female saint, it was probably sacred before Christianity imposed itself on the culture. While some avoid collecting holy water from a Christian church, even the cunning folk and wise women understood its significance and often used the water in folk magic. If this doesn't align with your path, seek out a holy person of your faith or path to bless the water.

If you don't have access to water from a sacred source, you can always use silvered water, which is created by placing a silver coin (or other piece of silver) in a bowl and praying over it. Hag stones or holy stones were traditionally used in Scotland for this purpose as well. You can also add pinches of salt or pieces of quartz or another power stone to the water and then pray over it. And if you are in need of a really powerful punch, use all three and then set the water out under the light of a Full Moon, making sure that the moonlight is reflected on the water. If you have to, walk around and tilt the bowl until you can see the Moon's reflection. Allow the water to charge for at least nine breaths, or leave it overnight for added power.

Sacred Vessels

Sacred vessels are central to a water priestess' practice. Throughout history—in literature, mythology, and archeology—we find numerous accounts of sacred women, the divine feminine, and goddesses who are associated with water. Many of them are depicted holding a vessel, gazing into water, bathing in water, or being birthed from the sea. There is no doubt that water is deeply connected with the sacred feminine and, for this reason, many esoteric and occult traditions associated the element of Water with the feminine. However, all genders can be represented by water, because it is fluid and ever-changing, and takes on the form of the vessel that contains it.

Because you will most often use and interact with water in its liquid form, you will need a vessel to contain it during your workings. But the purpose of a sacred vessel is more than practical. It also carries deep symbolism and potent magic. The sacred vessel, which is by nature concave, physically represents the female vulva, but also symbolizes the goddess, the divine feminine, the chalice, and the Holy Grail. It can also function as a scrying vessel, the womb of magic, or a holy woman, and can represent every body of water on the planet. It holds magic; it is the place of creation and gestation.

Many types of vessels can be used as sacred vessels, and many examples of them have been found at ancient sites. For example, the cup and ring marks found on megalithic stones in Ireland and Europe, and along the Atlantic seaboard, are believed to have had important significance for Neolithic people and are today considered pre-historic art. These markings consist of cup-like depressions carved into the stone that are often accompanied by rings or other marks that resemble lines. The depressions are large enough to collect rain water. Long after their original significance was lost, they became important in folk water magic, especially in conjunction with collecting dew or rain, which was believed to have magical and curative properties.

Perhaps the best known example of a sacred vessel is the Holy Grail, the chalice said to hold the blood of Christ (see chapter 2). This myth merges with Arthurian legends in Glastonbury, where the sacred vessel is represented by the Chalice Well and the well maidens' golden cups. Here, it expresses the divine feminine, whose body is one with the land. We find other examples of sacred vessels in the area as well—the ritual vessel displayed in the Lake Village Museum (Glastonbury), the numerous sacred wells within the Somerset landscape and the lakes that abound there, and in the legend of the Glastonbury blue bowl.

Cauldrons also appear as sacred vessels in Celtic myths—for instance, Cerridwen's cauldron, the Cauldron of Annwn, which is kindled by the breath of the nine maidens, the sacred bowl used by the Lady of the Fount, and the golden bowls carried by the well maidens. In the story of the Lady of the Fount, a faery woman who is perhaps also a priestess, uses a silver bowl, the water from the fount, and a marble slab to perform weather magic. The Gundestrup cauldron, another sacred vessel made of silver, was found in Denmark and dates from somewhere between 200 BCE and 300 CE.

Selecting and Consecrating Your Vessel

No matter what type of vessel you choose to use or have at hand, you must consecrate it and dedicate it to your sacred water work. You may choose a cauldron, a chalice, a bowl, a shell, or some other receptacle. If you find, over time, that you use different vessels for different workings, you must repeat this consecration and dedication each time you introduce a new vessel to your practice. I use several different bowls–black for scrying on the Dark Moon and during waning moons, and silver, or sometimes glass or crystal, for scrying on the Full Moon or for making Full Moon water (see chapter 6). Shells are wonderful to use as vessels in ceremonies or rituals, and are particularly appropriate if you are working near the sea. I use a little clay bowl to collect dew, and a blue ceramic bowl for holding water in cleansing and healing rituals. I also have bowls that have been imbued with magical intent for making flower and gem elixirs (see page 102), and wooden bowls for holding dried ingredients.

Selecting your first vessel may be easy or difficult, depending on how you feel about your practice. Perhaps you have an heirloom you want to use; or perhaps you just recently came across a vessel that attracted you. No matter what type of vessel you choose–vintage or brand new–the process will be the same for clearing its energy and consecrating it to sacred purpose. And make sure you choose a vessel that is safe to drink from as well.

Exercise: Cleansing the Vessel

Before consecrating a vessel or any item, you must first clear its energy. You can do this by making a cleansing wash out of herbs, but there are many other ways to clear energy as well. If you already have a method that works for you, feel free to use it. This exercise uses a magical wash to clear the vessel. In the next exercise, you will use holy or sacred water to consecrate it.

To clear the energy, you will use nettle, St. John's wort, and vervain. St. John's wort, which is also called "chase the devil," removes all negative energy; nettle provides protective energy; vervain purifies. Combine these with water in the vessel you are about to clear. The mixture will start to work immediately; you can finish the process with a spoken charm or prayer.

Begin by placing three pinches of each herb, dried or fresh, into the vessel. If you have access to sacred spring water, use it; if you don't, regular spring water is best. Heat the water until it is just below boiling. (Be sure your vessel can handle the hot water.) Then pour the water into the vessel, filling it about half full, and leave it out overnight. In the morning, strain the herbs and put the water in a jar. Select a clean white cloth, preferably a natural fabric, and use it to wash the sacred vessel by dipping the cloth into the herbal water and cleaning it as you would any dish. Don't rinse it; just allow the herbal water to stay on its surfaces. Then recite this prayer:

Prayer to Cleanse a Sacred Vessel

With the power of Water
I cleanse this vessel;
With the power of these sacred herbs
I cleanse this vessel;
With my spirit and my will,
I cleanse this vessel.

Exercise: Consecrating the Vessel

Consecration is the act of dedicating an item to a sacred purpose in a focused and mindful manner. When you consecrate your sacred vessel, you induct it into your array of ritual tools as a sacred container for the divinity of water and devote it to the highest purpose of serving the sacred waters. To do this, you will need either fresh clean spring water or Full Moon water that you have created yourself (see chapter 6).

When ready, place the water and your vessel on your altar or shrine. Dip your fingers into the water and splash it onto the vessel over and over till the entire vessel has been touched with the holy water. As you do this, recite the prayer below. I have included a blank space for the name of the deity you are working with, although you may not yet know who this is.

Prayer to Consecrate a Sacred Vessel

I dedicate this vessel to the water, the water spirits, and the goddess

_____.

I dedicate this vessel to my work and to serve my will.

I consecrate this vessel to the magic and mystery of the waters.

I consecrate this vessel with the power of the water.

From this moment forth, this bowl becomes the sacred vessel;

This sacred vessel is now cleansed, consecrated, and dedicated to the goddess.

May this serve the highest good and provide healing and service to the water.

So mote it be.

This sacred vessel will now serve you in all your future work, and you can use it to cleanse your water shrine and dedicate your temple.

Exercise: Becoming the Vessel

For many, sacred vessels have significance far beyond mythical or magical cups and actually contain special powers. Many see the sacred vessel or chalice as the divine feminine, or sacred woman. This ritual will help you become the sacred vessel, so you can use it to hold your own sacred waters and as a place from which you can work as a divine priestess. Full Moon water (see chapter 6) made with water from a sacred well is best for this ritual, but you can use sea water if you identify more as a sea priestess. Regular spring water can be used as well.

For this ritual, you will need:

- Quartz point

- Rosemary

- A sacred vessel

- Sacred water

- Salt

- White dress

- White roses

You may also want to have tarot or oracle cards on hand, and your Book of Tides.

Begin by taking a purification bath prepared with the following items:

- 3 pinches of salt

- 3 white roses

- 3 pinches of rosemary (or rosemary essence if you are pregnant)

When you are finished with the bath, dress all in white and go to your temple space. Place your sacred vessel in the center of your working space, altar, or shrine and pour your selected sacred water into it. Stand or kneel before your space and place the quartz point into the water. Stir the water nine times sunwise while chanting:

Waters of the Moon, come into me
So I can become a sacred vessel for thee.

Remove the quartz point and place it aside. Then dip your finger into the water and anoint your Third Eye, saying:

Sacred waters, I honor thee;
Open my Third Eye so I may see.

Anoint your heart center, saying:

Sacred waters divine,
I offer you this heart of mine.

Dip the quartz point into the water and draw the alchemical symbol for water on each of your open palms, saying:

May these hands work with the will of the great water spirit.

Take three sips of the remaining water, saying:

Sacred waters, flow through me;
Forever a sacred vessel I will be. (If you are working with sea water,
skip this step.)

Pour the remaining water over your head, saying:

Sacred waters that touch my soul,
A sacred vessel I am now, perfect and whole.

Sit back and move into a meditative space. Allow yourself to commune with the spirit of water and, if you get any messages or visions, be sure to record them in your Book of Tides.

Water's Lessons

There are many types of water you can use in your priestess practice, either to collect and use in rituals in your water temple, or to commune with at the source. Water can appear in so many forms, but, no matter its source, you can connect with it, learn from it, and share in its wisdom. If possible, try connecting with the water by touching it, holding it, or sitting next to it. You may also find that different locations in or around the same body of water have different energies and different lessons to teach. You may enjoy one particularly body of water more, or connect with one more easily than another. In my experience, waterfalls seem to have individual personalities, while larger bodies of water seem to give more than one message—perhaps because multiple types of spirits dwell within them, like merfolk, nymphs, and other water spirits. You can also learn a lot from the animals that live in the water.

When connecting with water, you may also receive the unsettling messages of pollution, death, or silence. If you do, don't worry. Silence may just mean that the water is not interested in connecting with you, or that it is not spiritually healthy or is in desperate need of cleansing or healing. Pollution is a worldwide problem, and I have learned that many water spirts ask for help in cleansing the water and healing it.

Each type of water has its own powers and properties:

- Brackish water holds the power of duality. This is less about polarity and more about the merging of the fresh and salt waters. If you can sit near brackish water, consider meditating there. What message does this brackish or grey water have for you? How can you integrate all the parts of yourself to create your own unique message?

- Creek water teaches a gentle lesson. The soft bubbles of these magical brooks can speak directly to your soul, giving a message of cleansing, healing, and even death.

- Dew is a renewable resource that you can collect using a bowl and a clean, thin wet rag. I personally use a vintage handkerchief with a lace border. I wipe dewy leaves with the rag until it is quite wet, then I wring it out into a bowl. This can be a long and painstaking process, but it can also serve as a moving meditation. The rhythmic collecting of the dew can bring you into a meditative or lightly altered state of mind where you can begin to commune with the plant spirit and learn the lessons it has to teach you through the dew.

- Floral and gem waters each have a different message. Keep a log of the lessons you learn from them, as each will be different and speak only to you. Be sure to write each one down to enhance your own learning and progress. Adds these notes to your Book of Tides and use them to create recipes for ritual waters, healing baths, and many other rites that you incorporate into your own practice or that you use in water workings for others.

- Hail can teach you the lessons of pain and hard decisions. The next time it hails, collect some and hold it in your palm as it melts. What message does it teach you?

- Ice teaches you the lesson of standing still and coldness, and how these can work for you on your own path.

- Lake water is still and reflective, it is often fed by springs and so has properties similar to spring water.

- Marsh water is filled with all sorts of life. It differs from swamp water in that it is usually not as dark and generally has fewer trees growing in it. Marsh water offers lessons of life and duality.

- Ocean water is vast and holds in its depths the power of dark and light, death and birth, beauty and destruction. It is a powerful force to reckon with. What message does the ocean have for you? Do you find different messages in different oceans? Is the Atlantic different from the Pacific? Meditate near each to find out.

- Pond water differs from lake water–especially if the ponds are man-made. While ponds may be newer bodies of water, however, they are still full of spirits, life, and lessons.

- Rain water is slightly different from storm water. It can bring messages of nourishment, of filling your cup, and of re-hydration. The next time it rains, sit outside and allow the drops to hit you; then contemplate the messages they have for you. Try collecting different types of rain in a bowl and holding it during meditation. Or meditate during a rainfall to connect with the water. Try this during thunderstorms, sun showers, or even windy rainstorms.

- River water offers messages about time and the wearing away of things that do not serve us. It can also deliver lessons about movement and change. Sit next to a river near you and listen to its messages. Does it speak to you of these things or of something else?

- Sacred water holds lessons that can be wondrous indeed. It is usually found at springs associated with a temple. If you have the opportunity to sit near or drink these waters, stop and listen to the sacred messages they may bring. This water has properties similar to holy water.

- Snow teaches the lesson of gentleness and slowing down, but also of beauty and light. Next time it snows, use your finger to carve a sacred symbol into it and meditate there. Record any message you receive–telepathically or otherwise. Try collecting snow in a bowl, bringing it into your temple space, and meditating with it. Sit in a light snowstorm or meditate near a window when there is a heavy snowfall.

- Spring water is alive with water spirits. It was once believed that every spring had a water faery or nymph that dwelled within. If you have the ability to swim in spring water or sit near a spring and meditate, try to contact the spirits of place, the *genii loci,* to see if they have any messages for you.

- Storm water delivers the message of aggression, but it can also teach you to wash away that which does not serve you. It may also teach you how to love the more aggressive aspects of yourself. It reminds you that your own darkness and anger can bring wisdom to your path.

- Swamp water is often home to poisonous plants, venomous reptiles, and spiders, so be careful when you connect with it. Contemplating these energies, however, can bring you a different perspective as a water priestess—a reminder that water can both heal and harm.

- Waterfalls teach lessons of beauty, of cleansing, of power, and of falling. They may even bring the energy of the Tower card from the tarot when it all comes crashing down around you. If you can sit near a waterfall, allow the spray to hit your face and bask in the message you receive.

Healing Gem Elixirs

A gem elixir is water that has been infused with the unseen vibrational properties, or energy signature, of a particular gemstone. While the instructions are for gem elixirs, the same process can be used for making floral elixirs that carry the properties of specific flowers and plants.

When you make gem elixirs, you must also consider the phases of the Moon in order to draw the correct energies down into the water. For a grounding water, align the creation of the elixir with the first- or third-quarter Moon, because, in these phases, the Moon is balanced between Full and Dark. These are the perfect times to work magic to balance your personal energies. Sometimes your magic may call for a solar infusion. In that case, use the Sun in a similar way to create an elixir, being sure to align with particular celestial events that suit your magical needs.

There are two methods for making elixirs—direct and indirect. Not all gem waters are safe to ingest, so be sure to do your research first and make sure the stones you use are clean and undyed. Use the direct method when you are sure that the elixir will be safe to drink. If you aren't sure, use the indirect method.

It is important to note that the *energies* of these gemstones are not toxic at all; they are beneficial. However, their mineral and metal structures make them vulnerable to leaching properties like lead into the water that may have an undesired physical effect when consumed or used topically. The indirect method still allows the subtle vibrations of the stone to imprint the water energetically, but protects you from possible negative consequences.

Exercise: Direct Method Elixir

In this method, you place the gemstone directly into the water. This is the best method for stones that are hard, that do not dissolve easily, and that do not leak biologically toxic compounds like lead into the water. Quartz, jasper, and agate are good examples of safe stones. Make sure your water comes from a clean source so it is safe. The water that you drink every day to stay hydrated is a good place to start.

To make this elixir, you will need:

- A clean glass or crystal bowl (this can be a simple glass mixing bowl)

- A stone that is safe to place in the water that will be ingested

- Clean drinkable water

If you choose, draw or paint a sigil or sacred symbol on the bowl, then pour in the water.

Charge your stone by holding it in your hands and focusing on your intent—for instance, to balance your energies and keep yourself grounded and connected to the earth. Place the stone inside the bowl. Hold your hands over the bowl and say any prayers or incantations that you want to add.

Cover the bowl with a clear lid or film that will keep the water from collecting dust, bugs, and other contaminants that float around our environment. Then leave it out under the light of the Moon all night. Retrieve it the next morning, remove the stone, and pour the water into a large mason jar. Be sure to label the jar and put it aside to use in your rituals.

Exercise: Indirect Method Elixir

Use this method when the stone you are using may not be safe. Black tourmaline is a good example of a stone that is very effective in removing negative energies and helping you stay calm and grounded, but black tourmaline water is not safe to ingest.

The indirect method is similar to the direct method, but requires a smaller bowl in addition to the larger one used in the Direct method. The smaller bowl creates a barrier between the water and the gemstone. A tall bowl with a rim higher than that of the larger bowl works best. Just make sure your larger bowl is big enough to hold the smaller bowl, as well as a good quantity of water.

Place the smaller, higher-rimmed bowl directly in the center of the larger bowl. Pour your water into the larger bowl, being sure not to get any in the smaller bowl. Stop pouring when the water reaches one to two inches from the top of the smaller bowl.

Place your stone into the small dry bowl. Cover both bowls with a lid or clear film and place them out under the Moon during an appropriate phase. Retrieve the bowls the next morning and remove the smaller bowl containing the stone. Then pour the water from the larger bowl into a mason jar and cover tightly. Label the jar and put the elixir aside to use in your rituals and workings.

Chapter 5

THE HEALING POWER OF WATER

Since the beginning of time, water has been used to nourish our bodies and give us life. It has been revered for its magical powers, but also for its healing properties. Spring water is often associated with healing and, in some cases, the springs or wells considered sacred in fact contained minerals that helped individuals recover from medical conditions. For example, springs and wells heavy in iron can cure anemia, while springs that contain calcium can heal certain skin conditions and improve circulation. In addition, trees are often associated with sacred wells, merging the healing properties of the water with their own power, spirit, and medicinal qualities. Today, these practices have become less common, because not all water is safe to drink.

Folk wisdom tells us that spring water combined with saffron was used as an eye wash to cure pain. Combined with prayer, the water caused the pain from the eye to pass into the water, then into the saffron, and finally out of the saffron and into the earth. The earth spirit devoured the saffron soaked in spring water and thus removed the eye pain. You can use this folk remedy as a model for your own work by calling on a plant spirit to work together with the water spirits and the energy of the earth to heal. In this kind of healing or cleansing work, the negative energies are held in the water and must be disposed of away from the subject.

Of course, you should always consult your doctor if you are suffering from a medical condition. Please remember that, unless you are a licensed therapist, certified herbalist, certified Reiki practitioner, licensed massage therapist, or any other type of body worker, it is important to let your clients know that your water healing sessions are for the spirit and the soul, and do not take the place of medical diagnosis or treatment. These practices do, however, work quite well alongside more traditional methods, and you can act as a spiritual healer for someone who has sought healing from other practitioners as well. We can still find the sacred powers in the waters and bask in their healing flow.

But this all brings us to consider: If sacred waters can have a physical impact on the body, can they not also have an impact on the spirit? What implications might these healing waters have for the dark parts of our souls? Could their divine essence provide the healing power we need? The ocean, for instance, has often been associated with emotional healing. We feel drawn to the sea, feel better when we inhale the salty air, and relax as we float gently on the waves.

While it is true that there is no specific pattern to ocean waves, they do often come in groups or clusters. Water collected from the ninth wave in one of these clusters was considered to have special healing properties, and it was believed that the Otherworld existed in a realm beyond the ninth wave. Legend tells us that the ninth wave will be slightly bigger than the previous eight, and that, if you stand in or collect water from this wave, it can be used in healing rituals and magical work.

When you collect water from the ocean or a running source like a river, be sure to collect with the flow of the water. Sink your vessel all the way into the water, rather than placing it with the opening facing against the current and allowing the water to flow into it. Gathering water against the current puts you in opposition to the force of the water and not in harmony with it. The Romani believe that, if you collect water against the current, it displeases the water spirits, so be careful how you gather your water.

It is also important to note that, although most water can be used in healing rituals and for topical anointing, not all water can be drunk directly from the source. There are exceptions of course, and most of those are sacred springs that have been tested and to which many make pilgrimages. Be mindful of this when collecting your water. You may want to invest in a testing kit to be sure that the water you offer in your rituals is safe to ingest. And be very clear with others as to whether or not your water is drinkable.

Healing with Water

As a water priestess, you make water the central tool of your practice in the healing and energetic arts. You may conduct your sessions in a special pool, at a particular spring or thermal spa, or in a water temple you have created in a room where you work water Reiki on clients. You may even work with clients out in the community. You may choose to have a moving temple and work with community members wherever you are able to set up a shrine for water rituals and healing. Or you may offer aura-cleansing sessions in which you asperge a sacred energy space with holy water and cleanse people by removing impurities while leading them on a vision journey to purify their souls in the other realms. In fact, water healing rituals will be different for each practitioner, as well as for each person being healed. The following are some examples.

I use Reiki in my personal practice. When I greet my clients, I offer them a Reiki or gem water, usually charged with healing intent. You can also start by giving them tea. Bring them to the space where the water healing session will take place, and allow them to talk if it makes them feel comfortable. I usually have my clients sit, however a massage table also works well. If you ask them to sit, make sure that there is enough room for you to walk around them in a circle.

I start with a casual phrase like, "So what's going on?" or "Glad you made it; so why are you here today?" This usually prompts them to pour out their hearts. I never make promises, however; I just nod and listen, or respond with something like:

"That is awful. I am so sorry you had to go through that. Let's see if we can remove some of that negative energy so your own can flow more freely again." Sometimes clients don't talk at all, so I just smile and tell them that I don't need any more information and that the water and water spirits will do all the work. Remember: The water will flow where it is supposed to, and you must trust that it will.

Exercise: Clearing Negative Energies

Light blue healing candles and pick out a stone for your clients to hold. I use blue lace agate for calming, smokey quartz for grounding, or blue calcite to facilitate healing. This gives clients something to hold onto and makes them feel more secure rather than nervous or anxious. Just be sure to cleanse the stone with sacred water when you are finished.

Begin with a tonal technique like drumming to help ground clients and set the mood. Use whatever technique works for you. Then begin to walk around the client while dipping a fresh herb bundle or fresh sprigs into the water and flicking or shaking them so that the water lands on him or her. Rosemary and St. John's wort both work well in clearing energy. Use fresh sprigs and herbs rather than dried plants if you can. Make sure to let your clients know before you do any of this, as many keep their eyes closed and the cool water can be jarring. Then start to feel for energies that need to be repaired, healed, closed, or opened, and do the energy work accordingly.

If clients consent to be touched, you can use water to draw healing symbols on their bodies in weak areas. You can use runes, sigils, and other sacred symbols as well. The head, feet, palms, and heart center tend to be good places to work. If you need to do this in a clothed or sensitive area, use the herbs to flick more water on the area and draw the symbol in the air several inches away from the body. You can work in the aura space and on the body; both are efficient. The important thing is to make your clients comfortable and not violate their space or make them uneasy. Healing is a delicate art and you need to make your clients as relaxed as possible and feeling as safe as possible to get the best results.

If you are not an energy practitioner, but still want to perform healing rituals, you can begin by praying or chanting. Craft a healing prayer that asks the water to bring its healing energies, or call on the power of a water goddess or mermaid guide to bring healing water energies from the unseen realms. Prayers, incantations, and repetitive chants have been used across many cultures throughout human history as a means to change the energies in spaces, as well as in the body, the spirit, and the aura.

Using sacred water, begin to bless the body as you pray or chant. Anoint the brow and say:

May your visions be blessed and true.

Anoint the eyelids and say:

May your sight be blessed with beauty and love.

Anoint the lips and say:

May your words be blessed and true.

Anoint the heart center and say:

May your heart be blessed and rest in divine love.

Anoint the palms and say:

May all that you touch be blessed and sacred.

Anoint the feet and say:

May everywhere you travel be the right and true path.

Anoint the crown and say:

May all that is from here below be blessed and full of grace, love, and abundance.

With the power of these sacred waters, this is my will; so will it be!

When the session is over, use a tonal technique like drumming to let clients know that the session is coming to an end. Then bring them through a very short visualization in which they travel through a beautiful waterfall and cleanse themselves on the astral plane. Lead them back with your voice to the present place and time, and ask them to begin to move slowly, wiggling their toes and fingers and slowly opening their eyes. Ask how they are feeling and then offer them some water–perhaps an energizing type of water or something more grounding, depending on their state when they emerge from the session.

Sometimes I prepare waters for clients to drink before bed, or to bathe in that evening to help continue the process. I caution them to take care for the rest of the day, to drink lots of water, and to rest, as healing and energy work tend to affect people differently. The power of water brings healing, but can also amplify and enhance change and transformation during a healing session.

Creating Healing Waters

As a water priestess, you will work with a variety of water on a daily basis. And while you will probably find that you have a favorite type or body of water, you may also want to create waters you can use specifically for healing. You may want to create water that is specifically attuned to healing or love so you can perform healing rituals for others. Or you may want to create gem-empowered waters to bring with you as an offering when you do a river blessing (see chapter 4). This is a beautiful way to enhance the energy of the environment without harming it.

We know that water absorbs energy, holds memory, and has a particular vibrational signature that depends on how and with what it is being used. This tells us why water, used in conjunction with other sacred vibratory tools, can enhance your energetic healing work. Early this century, a Japanese scientist named Masaru Emoto published several books documenting his work with water in various forms. For years, he recorded the changing appearance of water crystals from different sources as they were exposed to different types of sounds and music. He documented his work with hundreds of photos that compared water crystals from a sacred source with crystals from a polluted source. Then he took his research one step further and subjected the water crystals to different types of music—classical, rock, and heavy metal. He also subjected some crystals to different words and other sounds, and then compared what they looked like. He found that crystals formed under conditions of cleanliness and harmony were themselves beautiful and harmonious, whereas crystals formed under conditions of pollution and stress were shrunken and distorted.

Magicians understand the results of Dr. Emoto's research as part of the theory of vibration that drives their craft. And while we may not all be equipped to document our experiences with how the water changes physically, we can apply Emoto's conclusions to our own water craft. If water can change its structure based on the vibrations of sounds and music, then it is reasonable to assume that the water will pick up and absorb the vibrations of stones and herbs. This is one of the reasons it is very important to research your stones before plopping them into the water; you don't want to drink toxins or gulp down water contaminated by a dangerous additive. Selenite, for example, is water soluble, but produces thin sharp shards that are very dangerous to drink. Always research your stones before making gem elixirs. If you are unsure of their properties, use the indirect method found in chapter 4.

There are many ways to obtain or create healing waters. You can gather it from a sacred spring or the ocean, or you can use old folk methods, like adding a silver coin or hag stone to the water. You can add salt to water, or bless it with sound, prayer, or incantations, or charge it with various crystals, moonlight, sunlight, or sacred herbs. Water absorbs and enhances the energy of whatever is added to it.

You can charge water with the energies of the Moon or the Sun by pouring it into a sacred vessel and placing it in direct light from the Sun or Moon (see chapter 6), then reciting the water blessing prayer below. You can also do this with crystals and herbs. Just make sure you stir the water nine times as you recite the prayer. Stir sunwise to bring in positive energies and counter-sunwise to banish or to connect with shadow energies.

Sound is also a good way to create healing water, and there are a few different ways to do this. You can simply start with your voice and tone the water with vowel sounds. You can also use color, introducing different-colored jars or bottles to help charge the water with the healing energies of the color.

Water Blessing Prayer

Nine times the water spins;
This sacred prayer I now begin.
From the heavens above,
Bless this water with healing and love.
From the deep below,
Empower these waters with healing flow.
With my breath and with my will,
I calm these waters and make them still.
This is my will, so mote it be,
Healing waters three times three.

Exercise: Shell Healing Ritual

Most of us have a collection of shells, even if some of them are old or broken. No matter how small your collection is, you can cleanse and consecrate it using the methods presented in this book. When choosing your method, however, keep in mind that this ritual is intended for healing. You can choose from the associations given above, or you can draw on your own correspondences. Just be sure to be consistent.

You will need a photo of yourself or the person you are healing. If you can't obtain a photo, write out the person's full name, date of birth, and any appropriate sacred symbol on a piece of paper and use it as a taglock instead of the photo. If you can get nail clippings, clothing, or hair from the target person, all the better. If you are working for yourself, use your own hair or saliva, or a photo, or even your breath.

Place the photo or taglock in the center of your working space, then use the consecrated shells to create a healing grid or design surrounding it. This can be a simple circle, or you can use other sacred geometric shapes. It can be symmetrical or not. The important thing is that your artistic creation be done with intent, mindfulness, and purpose.

As you create this design, charge each shell with intent the same way you would your other magical items. Once the grid is laid out, anoint five blue candles with sea water and place them around the design, evenly spaced. As you light each candle, starting at the top and working sunwise, ask the water spirits to bring healing. When you are finished, use your ritual knife, your wand, your finger, a quartz crystal, or even a special healing shell to activate the grid by drawing connecting lines of energy from candle to candle. This will create a pentagram over the shell healing grid. Be sure to check, recharge, and tend the grid periodically. Note any changes that may take place. The ritual is complete when the candles have burned down completely. When you are finished, or when you believe the healing has taken place, deconstruct the grid and cleanse the shells for future use.

Exercise: Healing Waters Dream Pillow

A dream pillow is a fantastic way to use herbs and plants to aid you during your dream states. You will create this dream pillow with the intention of receiving messages for healing while dreaming. Instead of sewing a pillow, you may wish to create small sachets. You can paint, sew, or embroider the symbol you choose on the pillow or sachets. You don't have to use every item listed below, but be sure to use at least three of them for the sake of potency.

- 2 pieces of white or light-blue cloth cut into squares of about 6 inches

- Needle and thread (perhaps blue or white)

- 1 pyritized ammonite to protect your dreamscape

- Mugwort to open the dream portal

- Lavender for rest and protection

- Eyebright to enhance your second sight

- Rose petals to connect to the divine feminine in the waters, in the universe, and in you

- Sandalwood to purify the energies around you

- Heather to protect you as you sleep

- Sacred water from a dreaming spring if possible; if not, spring or ocean water

- Dark-blue thread, string, paint, or other choice for creating sacred symbols on the fabric

Begin by gathering your fabric and healing water. Use the water blessing prayer or one of your choice to consecrate it. On the night before the Full Moon, place both pieces of fabric in the water and let it soak for three hours. During this time, you can pray over it, chant, drum, dance, or do anything you like to help add additional intent and energy to the process. Then wring the fabric out and allow it to dry completely. Keep a tiny bit of the water for the final process.

On one piece of the dried fabric, draw, paint, or embroider a blue alchemical symbol for water–a triangle with the point facing down. On the second piece of fabric, put your own sacred symbol (see chapter 9). If you are working for others, place their names on the second piece.

Place the two pieces of fabric together with the symbols on the inside. Sew three of the four sides closed, forming a small pouch, then turn the pouch inside out so the symbols are on the outside. Use the open side to stuff the pillow with the items you chose from the list and anything else you want to add. Be sure to tuck the ammonite or any other stone into the center so the pillow remains soft. Then sew up the remaining side.

Bless the final knot that closes the pillow with a single drop of the water you prepared, sealing the project as completed, blessed, and ready for use. Place a second drop on the side with the alchemical triangle, blessing the pillow and activating it as a tool to bring dreams toward you easily. Anoint the final side, where you placed your own sacred symbol, with a third and final drop, blessing the symbol and connecting it to your energy. When you are finished, place the pillow or sachet on your bed, keep it in your magic room, or pull it out for dream meditation rituals. Be sure to record any dreams or interesting experiences you have in your Book of Tides.

Healing Water Deities and Spirits

There are many water deities and many more water spirits who are associated with healing. What follows is by no means a complete list, but just examples of some of each. There are numerous ways to contact these deities. You may consult spirit dictionaries or watery-themed oracle decks to learn more about the many types of water spirits out there. When you work in service to the water or in healing, you will no doubt come into contact with these entities and forge relationships with them—or perhaps even dedicate yourself to a particular one. When choosing the deities with whom you will work, be sure to research the practices, cultures, and taboos surrounding them.

Dedicating yourself to a deity is a very serious process. It is not something you can or should decide on overnight, even if the goddess or spirit you are approaching has been with you for your entire life. As with any friendship or courtship, you first must meet the deity, then begin to communicate to establish a working relationship. Not all spirits or deities are benevolent, nor will they all have your best interests at heart. So get to know these entities and make sure that you can not only work with them, but form a trusting relationship with them and be able to fulfill your obligations.

Once you dedicate yourself to a deity, you must follow through on your commitments. Take this very seriously, because there are often consequences for failure when it comes to working with certain spirits. Most goddesses have a gentle and forgiving nature, but this is not true of all of them. Merfolk and other finfolk are known to be temperamental and can also be mischievous. They tend to anger when they are let down and when the water is harmed. They have been known to grant wishes and give treasures, but also to steal souls and drown people. I don't mean to scare you, just caution you to move through these deep waters with wisdom and with a clear understanding of the Otherworld and the realm of the spiritual waters. If you are drawn to a particular goddess, you can use the dedication ritual in chapter 9 to consecrate yourself to her service, but first consider very seriously whether or not this is for you.

When healing with water, you can also call upon wisdom and guidance from ancestral water priestesses. Whether practicing dreamwork, divination, scrying, or good old-fashioned summoning, they can help to guide you. You may want to work with Dion Fortune, author of the influential novel *Sea Priestess*. Or perhaps you are drawn to the Irish Bandruí, the Norse Volva, the priestess of the Isle of Sena, or other Galician priestesses. You can turn to the oracle at Delphi or the priestesses of Avalon—Vivian or Morgan la Fae or other priestesses who dwell in Avalon.

While Morgan is an Avalonian priestess, many venerate her as a faery queen or goddess.

There have been many water deities across time and space, and existing in many different cultures, who have been associated with healing. In general, deities aligned with the sea have a powerful energy of transformation, healing, and authority and dominion over the animals and spirits that dwell in the ocean realms. Some are believed to be the personification of the ocean, while others are singular goddesses who reside in the seas. In other cases, they are born from the waves or are considered important nereids or nymphs.

Likewise, there are sacred springs all over the world associated with healing. Some flow with waters that heal through pure minerals, while others put forth healing through warmth. There are also some rare cases of springs that flow with poisonous waters. Because goddesses were often associated with these springs, it is not surprising that there are numerous temples built near, on, or next to these sites. This was especially true in the Middle Ages, when the sacred springs and wells were renamed for saints and churches were erected there.

The sacred temple at Bath is watched over by and dedicated to the goddess Sulis, called Sulis Minerva by the Romans. But Sulis was venerated in this location long before the Roman occupation of Britain. When her temple at Bath was excavated, they found it filled with requests for healing and with more than 100 curse tablets! Sulis's name means "Eye of the Sun" or "Goddess of the Gap," depending on the source, and she was traditionally associated with water and healing in both Celtic and Roman traditions. Her spring waters bubble up and into her temple, hot and steaming and packed full of minerals, reflecting her Sun aspect.

Coventina was another notable goddess connected with sacred springs who was worshiped by both the Celts and Romans for her healing powers. Her shrine, which is in Northumberland, has yielded up countless coins and pieces of jewelry, as well as two ritual masks of women. The story of Melusine, another complicated water goddess who takes on many different aspects, is also associated with several different sacred healing founts.

Finally, we cannot forget Brighid, probably the most famous goddess associated with the healing power of sacred springs. Brighid, who was later re-named St. Brigit in the Christian tradition, had many wells dedicated to her, the most famous being in Kildare, Ireland. Brighid, who is quite popular in modern Paganism, was honored as both a Fire and Water deity who brought healing. Later, as St. Brigit, she became associated with the sacred well and shares many common attributes with her predecessor. Today, many who honor the goddess and/or the saint visit her well on holy pilgrimages to honor the land and partake of the healing waters. During the feast of Imbolc, or Candlemas, people honor Brighid in her many forms by creating traditional straw crosses in her honor and asking for healing. Clooties, rags or strips of fabric dipped in the sacred well water, are often hung in trees with the hope that they will heal ailments, cure disease, and ease troubles of the heart. This practice aligns with many other sacred well traditions.

Rivers throughout the world have also been named for water deities and spirits. In fact, in antiquity, rivers were often seen as the very personification of a water goddess or nymph. Belisima is one such figure worshiped by the Celts. The River Boyne is named for the goddess Boann, while Ganga is a water goddess considered to be the personification of the Ganges. Likewise, Sequana is goddess of the Seine. In Egypt, Anuket is goddess of the Nile, while Nepthys is the goddess of all rivers and death. Many consider Isis to be a water goddess as well.

While the water is most often associated with the feminine and water goddesses, we would be remiss to exclude the vast number of water gods who have left their mark on cultures around the world. They are too numerous to detail here, as they could fill an entire volume by themselves. However, in case you are drawn to honor the masculine side of water and the fierce guardians and kings who preside over it, we will consider a few of them here.

The most notable, of course, is Poseidon, also known as Neptune, who is King of the Sea. According to Pausanias, a writer from the first century CE, he was connected with the oracle at Delphi. Pausanias describes in detail Poseidon's altar in the temple of Apollon. He is considered the protector of the sea and all aquatic creatures and is widely worshiped by seamen.

Other notable water gods include Triton, a Greek god who was the son of Poseidon and Amphitrite. He is considered to be a messenger of the sea, and is usually portrayed as a merman, with the upper body of a man and the lower body of a fish. Like Poseidon, he carried a trident, but also a shell that he could use as a horn to calm or stir the waves. Pontus, the son of Gaia, is a primeval god of the sea who was said to have been conceived without "the sweet union of love." He is the personification of the sea and much more. Together with Gaia (the Earth), he fathered Nereus, who is described as the old man of the sea. Nereus was the father of the nereids and the nerites, and lived with them in the Aegean sea.

In Celtic lore, Mannanan Mac Lir appears as a god of the sea, along with Gwyn ap Nudd, the faery king. Gwyn, while not originally a water god, is deeply connected to the White Spring in Glastonbury. Grannus is another Celtic god who is associated with sacred and thermal springs, as well as with healing. Lir and Nodens are two notable water gods, while boatmen like Barinthus and Charon appear as spirits who help those seeking the Otherworld through the liminal spaces and across the astral waters to Avalon and Hades. In Egypt, Khnum, Sobek, and Hapi were water gods associated with the Nile. In fact, every culture across time and space has boasted masculine water deities.

There are also numerous types of spirits who exist beyond the mundane world and work with water priestesses to heal and transform. These include nereids, undines, mermaids, selkies, nymphs, and lake ladies. It would take an entire volume to describe and name all the spirits that belong to the water and the lore and customs that surround them, so I have instead detailed the major categories and hope that they inspire you to dive deeper beneath the surface.

The nereids are the fifty daughters of Nereus, who was the son of Pontus and Gaia. They are considered to be sea nymphs. Unlike other nymphs and naiads–humanoid creatures who dwell in rivers and lakes–the nereids dwell primarily in oceans. Naiads are specifically known to preside over and protect fountains, sacred springs, and other sacred fresh-water sources. Water nymphs are similar beings, however they are not restricted to fresh water as the naiads are. Oceanids are the salt-water version of naiads. The Germans called these spirits *necks* or *nixies*, however it seems that water spirits in this classification were more commonly male.

Undines, first mentioned in classical texts by Paracelsus, are considered to be the elemental spirit, or the incarnation of the element of Water. They appear in many different forms, but are most often depicted in female human form. Later, undines were merged with nymphs and are often thought to be the same. As the personification of water, undines are associated with almost all bodies of water, but most often inhabit streams, brooks, springs, and fountains. They are rarely seen, although they are often glimpsed momentarily as shadows or ripples in the water. Rosicrucian legend tells us that an undine can acquire a soul by marrying a man who will be faithful to her.

There are numerous stories about supernatural water spirits who dwell in lakes. These are known as lake ladies. The Lady of the Lake from Arthurian lore is perhaps the best known, however there are stories that pre-date her. In Wales, these spirits are called Gwragedd Annwn, or wives of the underworld, and cows are sacred to them. A few other water spirits and goddesses–for example, Brighid–are also connected with cattle.

Mermaids are water spirits who appear across many cultures and traditions. They are found in both fresh and sea water and are usually depicted with a fish-like tail. There are stories of mermaids being stranded on shore and falling in love, tales of them being captured and held in lakes, and legends of them taking vengeance and guarding treasures. Sirens, though originally bird-like beings from Greek mythology, were known to lure sailors to their death by shipwreck with beautiful song. Now the word "siren" is almost synonymous with our classical image of a mermaid. Over all, these creatures have a complex nature.

Well maidens are spirits of the Faery–nymphs and ancestral priestess who once guarded the sacred wells and tended the sacred landscape. Again, the most famous of them comes from Arthurian lore in the story of the maidens with their golden bowls and the Wasteland. In yet another story of a well guardian, the Lady of the Fount, a faery woman, sprinkles water from a silver bowl onto a marble slab to call forth rain and thunder.

Exercise: Water Goddess Ritual for Healing

This is a healing ritual for emotional, spiritual, and energetic wounds. It is not a substitute for medical advice or help, and should not take the place of it. However, if you are working through some deep shadow issues, facilitating ritual and holding space for others to grieve, or working through trauma, this ritual can help.

For this exercise, you will need:

- Black or dark-colored bowl
- Blue and white candles
- Chalice
- Drinkable healing water of your choice
- Fresh white rose petals gently removed from the stem
- Healing altar
- Myrrh incense
- Paper cut into several small pieces and a pen
- Statue, image, or effigy of a healing water goddess

Begin by creating a water altar as described in chapter 4 and place the healing goddess of your choice in the center above the black or blue bowl you are using as your sacred vessel. Pour your healing water into the bowl. Use the rose petals to create your sacred space. Depending on what resonates with your intent, you can place the petals in a circle, in a spiral (with you sitting in the center), in a triangle with the point facing down, or in the shape of the Vesica Pisces as a representation of the yoni and the sacred feminine.

Once you have done this, sit in the center of your healing space and bring yourself into a meditative state. Call upon the goddess for healing with the chant on page 128. Remember: You are safe in your sacred space, cocooned by rose petals and the love of the healing waters.

Begin to focus on who or what you want to heal, allowing yourself to process and feel the pain, sorrow, grief, loss, betrayal, or any other negative emotion. As these come to your mind, write each down on a separate piece of paper. When you are finished contemplating each negative state, say:

You were once here, but no longer; my waters are full, healed, and clear.

Place each piece of paper in the dark-colored bowl. If you cry, allow your tears to fall into the bowl as well. Then pour some of your healing water over the negative emotions. As they absorb the water, begin to tear them apart, chanting:

Out, out, and away; you do not serve me today.

Once the paper has turned to mush, rinse your fingers with more of the sacred water and place the dark-colored bowl away from the altar. After the ritual is finished, take the negative-emotion paper soup and pour it into the earth.

Fill your chalice with the rest of the healing water, then hold it in your hands and bring yourself into a meditative state again, this time visualizing yourself as whole, as perfect, in your best form, and serving as a water priestess. What does this look like? What does it feel like? What positive attributes shine through? When you have a clear vision of this in your mind, charge the water with the visualization, pushing the energy out of your hands and into the chalice, and blowing your sacred breath of life onto the water. You can chant the vowel "U" (whose shape resembles a vessel) to align with water, or another sacred sound or word that represents you as a healed and perfect water priestess. When you have fully charged the water in the chalice, drink it and allow the healing to flow through you, binding up old wounds and energizing you toward your true and best form.

Water Goddess Healing Chant

Waters from the great below,
Help me heal and to grow.
Water Goddess, hear my plea;
Cleanse, heal, and purify me.
Remove these blocks and all this bane
So I may serve in wholeness again.

Chapter 6

THE WISDOM OF
TIME AND TIDES

*T*he power of the Moon over the waters is something that can't be denied. Menstruation cycles often align with the Moon's phases or, at the very least, mirror them in its waxing and waning as the uterus fills and empties. Even those who do not menstruate feel the pull of her energies and the power of her gravitational force as it affects the waters of this planet. And that includes the sacred vessel that is you, and the water contained within you.

The gravitational force of the Moon pulls and pushes the tides. Her celestial nature illuminates us like a cosmic mirror hanging in the sky, reflecting and shedding light on the darker aspects of ourselves. The tides connect the cosmic waters and the celestial energies of the sky and the heavens with the earthly realm of water, which each night reflects the beauty of the stars back to the heavens. What wisdom can we gain here? Why are there so many stories of star goddesses falling into the sea, or mermaids falling from the sky into the water to become divine beings? What can this tell us about the sacredness of water–this element that belongs, not only to this planet, but perhaps to the entire universe?

As the Moon affects the tides, so the tides affect us. You can use this knowledge in your magical work by aligning your rituals and magic to particular phases of the Moon. You can also align your work with the tides. Incoming tides are great for manifestation work; outgoing tides are great for banishing. Paying attention to the way the tides and the Moon behave can shed light upon the nature of water and on your own work as a water priestess. By aligning the Moon's cycles with the tides and merging these energies, you can further enhance the energy surrounding your workings. Work with the Full Moon at high tide for manifestation, empowerment, protection, and healing; or use the waning moon at low tide for banishing, releasing, removing, and reconciling shadow work. Likewise the Dark Moon at low tide brings energy to stillness, reflection, rejuvenation, planting seeds, and discovering shadow work. The waxing moon at high tide energizes rituals for manifestation, growth, healing, and prosperity, while work done during an eclipse can uncover that which is hidden.

Exercise: Charging Moon Water

You can create Moon water for use in your rituals and workings by placing pure water or water from a sacred source where it can absorb the Moon's energies. Different phases of the Moon will create waters with different properties. Create Full Moon water by placing your water in a silver, white, or clear glass bowl and setting the bowl out where it can capture reflections of light from a Full Moon. Let the light of the Moon shine directly onto the water to imbue it with Full Moon energies. You can also add plants with associations to the Moon or crystals like moonstone or quartz. Let the water absorb the energy from the Moon's reflected light overnight, then bottle it and save it for your Full Moon rituals and workings.

You can create Dark Moon water in a similar way, using a black bowl and setting it out during the night of a Dark Moon to soak up the dark, nocturnal energies of a moonless sky. When the water is fully charged, bottle it and save it for your Dark Moon ceremonies.

Exercise: Shapeshifting with the Lunar Cycles

The Moon in her spiritual nature is a shapeshifter. She not only shifts her shape and energies, but she also influences us differently as she changes. She is the mistress of illusion; she is the serpent who sheds its skin only to be reborn again and again. The Moon does cycle through phases each month. But it is the play of light and shadow–like the nature of water–that creates the illusion that she actually changes shape each month. You can use the power of this illusion in your workings.

Work with the cycles of the Moon through ritual bathing to manifest positive changes in yourself. Starting directly after the Full Moon, bathe every day while meditating on that which you want to shed. Finish each bath by visualizing it going down the drain with the water. At the New Moon, take a ritual bath and sit in the stillness of your becoming. Contemplate what you want your new shape to look like. What traits did you leave behind that made way for this fresh new shape? What will it reflect back out into the world? For the next few days, as the Moon begins to wax to Full, spend time in ritual contemplation while relaxing in the waters. Consider what you want to grow and how you want to nourish it. You can repeat this cycle whenever you like or do several cycles in a row for things that are harder to shed and manifest.

Exercise: Full Moon Ritual Bath

Perform this ritual on the Full Moon. At least one month before your ritual, create some Full Moon water, as described on page 134. You may also want to pick up a rainbow moonstone or regular moonstone or a small quartz ball and infuse your Moon water with these crystals. Once you have created the water, bottle it, label it, and store it in a dark place until the following Full Moon.

When the day of your bath arrives, fill your tub with the desired amount of water, sea salts, white rose petals, and nine drops of the Full Moon water. Once you have prepared the bath, focus on the energies of the Moon and recite the Charge of the Water Priestess to pull down the power of the Moon. This bath will help you charge your own batteries and remind you of who you are and why you are here.

Charge of the Water Priestess

Listen to the words of the water,
She who, of old, was called by many names
And whose form is vast and many;
She who is the womb of creation,
The fluid life force of the earth,
The ever-present longing,
The joyous song of our soul.

This world is ever-changing as my tides will always be;
You must flow fast like the river,
Turn your sacred attention to me;
I call you now, water priestess.
Arise and bear my call.

My agenda is to heal you,
My powers filled with might.
I call you to the waters so you can become the light.
Like the rushing river, I have worn away your pain;
I have stripped you layer by layer like a fierce downpour of rain.
I heal your wounds in the water in the holiness of my name.
While my nature is of shadow and beautiful joy's light,
It takes a powerful person to join in water's fight.
My waters are polluted; I'm wounded every single day.
You are a powerful priestess who can show others the sacred way.
Once I was sacred, temples built worldwide in my name;
I held the magic waters, healing every day the same.

Your second sight, a gift from me,
I bring you visions to do your work.
Listen! These are key!
I will guide you through the waters,
Still lake and raging sea;
I will lead you through the wasteland
So you can restore my sacred waters once again.
Sacred waters shall ever be.
Arise, water priestess, and hear my desperate call;
Return to the temple, build it, and stand proud and tall.
You are my powerful conduit, from river bank to ocean sands;
Feel the power of water flow forth from your hands.
Heal the earth, clean the waters, and protect her sacred lands.

Your duty to the water will surely change the world.
Do not despair in the darkness that we so often see;
Turn your sacred work and direct it at me.
Together we are many; together we can make change.
I pour from you, my water vessel, out onto the sacred land.
The change is here and starts between your hands.

Exercise: Full Moon Water Protection Ritual

This ritual should be done on site at your chosen shore. On the
Full Moon, venture to the water's edge with a sound tool of
some type—drum, singing bowl, your voice, etc. Begin by setting
your ritual, then find a large strong stick and draw a circle in
the sand around the ritual space. Draw nine tridents around the
circle, with their base at the edge of the circle and their points

forking out toward the water. If there is no sand, you can construct the symbols from sticks or small stones. You may have to use only three or six tridents; just make sure you use a multiple of three. Tridents are known as tools of protection and carry an interesting correlation with the Norse rune Algiz, which looks like a pitchfork or trident with three prongs branching outward. Not surprisingly, the trident is also associated with protection for many modern practitioners.

When your ritual space is prepared, begin to create a sacred sound. Regardless of which instrument you use, begin to flow with it. As the sound begins to vibrate, let your eyes drift shut and visualize a beautiful silvery-blue light pulsing in the instrument. Let the light pour out like water, strengthening with each sound. Let it pour from you and into the circle. See the circle, and then the tridents, begin to glow blue. Once the entire space is glowing with beautiful light, see the liquid energy pour out from the tridents and flow toward the water. When the light reaches the water's edge, let it flow out to the horizon, and even farther, while continuing your tonal work. See each molecule of water become infused with this liquid light and its protective energy. When you are finished, slow your sound down and diminish the light flow until only drops remain. Do this whenever you feel that the water is in need of protection.

Exercise: Full Moon Water Healing Ritual

Begin by preparing any of the powerful water blends found in this book or use your own creation. Just be sure that the blend you use does not contain any oils or additives. A simple blend like spring water that has been charged with the reflection of the Moon and a quartz crystal works well. Then journey to the body of water you are going to work with, bringing your sacred vessel with you. When you arrive, spend time near the water's edge. If it is a river, gather several rocks to create a healing sigil, a spiral, or the Reiki power symbol *Cho Ku Rei*. If it is the ocean, draw in the sand; if it is a lake, you may be able to do a little of both, using broken branches, driftwood, and long grass to create symbols if rocks and sand are not available.

Once you have prepared the shore, set your sacred vessel near the water's edge and pour your water into it. Place your hands over the water and see it begin to glow with a beautiful healing blue light. Let this water sparkle and glitter; see it in your mind's eye, flowing out of the bowl, over the sand or rocks, and into the water. As it reaches the shore, see the blue light begin to spread throughout the water, mingling with it and flowing along with the body of water. Then see the waters spreading all over the world, wherever the waters flow.

When you are finished, stand up, take your bowl to the water, and empty it there. Visualize the same blue light flowing out into the waters and throughout the world. Now place your hands in the body of water and, one last time, visualize that beautiful blue healing light being sent into the water and throughout the world.

Exercise: Dark Moon Ritual Bath with Persephone

This healing ritual is best done in the very last stages of the waning moon, when the Moon is a tiny sliver or not visible at all. The bath calls on the energies of the underworld and the queen of darkness and duality, whose realms lay across the River Stix. It is here, in this stillness, that you can pause, take a deep look at your wounds, and heal your emotional trauma and pain.

For the ritual, you will need black bath salts (not salts made with ash or dye), large garnet pieces to decorate the sides of the tub, dark burgundy and black candles, and a large pomegranate. Cut the pomegranate into quarters or eighths and place them in a dish. Keep the juice in the dish to use later. You will also need a quartz point and a small dish of sacred, holy, or healing water.

Fill the tub with the salts and water, light the candles, and step in. Begin to meditate and contemplate the trauma you want to heal. As you focus on it and feelings begin to well up inside, allow yourself to feel and process them. As anger and unbearable feelings begin to surface, let your tears fall–cry, scream, and expel all the pent-up energy. Then take a slice of pomegranate and squeeze it over the water, feeling the pomegranate jewels pop and seeing the red juice flow like blood from your hands and into the bath. As you do this, call on Persephone; ask for her healing guidance; ask her to heal you so that your heart will no longer bleed. If there is a psychical location on your body where you hold the energy of this trauma, or where you perhaps experienced this trauma, squeeze another pomegranate section over that area. As you squeeze it, see the wound in your mind's eye; see it also bleeding and flowing. See it flowing out of you like blood from an open wound.

Do this until you feel that your emotions have been processed and you are feeling better. Then dip the quartz point into the healing water and hold it on the wounded area, visualizing the wound healing, seeing a healing sigil form in your mind's eye. When you are finished, thank Persephone and commune with her one last time. Then drain the bath. You will have to clean it quite well after this one! Place any pomegranate seeds that were not used in the ritual, a piece of garnet, and a little holy water on your outdoor shrine as votive offerings to Persephone.

Please note: This bath does not heal physical trauma. You should consult your doctor to guide you in healing physical ailments. This bath is specifically focused on the energetic, spiritual, and emotional healing of trauma. When you suffer through terrible incidents like abuse, addiction, separation, or loss–even if you try to process them in the moment and use healthy coping techniques–the energy of the event can build up and be stored in your energy field. For some, trauma can manifest as a mysterious ailment localized to a particular part of the body. Like the cord-cutting bath on page 142, you can repeat this ritual often, until you no longer think about or have emotions that trigger the trauma. In fact, you may wish to follow the cord-cutting bath with this one to seal in the healing energies.

Exercise: Dark Moon Bath to Fill Your Cup

This bath is all about filling your cup. When the Moon is dark and absent from the sky, it is a good time to stop, to withdraw, and to take time for yourself. You have to fill your own cup, your chalice, so that you can give to others. This is a necessary ritual that you can perform often to help you do that.

For this exercise, you will need:

- Fresh organic rose petals in a color of your choice

- 3 pinches dried catnip

- 3 pinches dried hyssop

- 1 pinch valerian

- 1 cup Epsom salts

- Rose quartz

- White or blue candles

Grind the catnip, hyssop, and valerian with a mortar and pestle and combine with the Epsom salts. Then charge the mixture with rose quartz. You can also place rose-quartz crystals on the side of the tub as you bathe, along with white or calming blue candles. Once the bath is prepared, add the rose petals and the herbal salt mixture. You can also add lavender and rosemary to create a delicious scent! Step in and relax.

When you are ready, visualize the most beautiful chalice you have ever seen. Imagine that this chalice is you—a representation of you in the astral realm. Now see a beautiful silver celestial rain coming down from the heavens and filling the chalice. See the reflection on the surface as each drop falls into the cup. What are these drops? How can they feed you and your energy? Can you see what is filling your cup? Is it a nourishing light? A beautiful energy?

Once you have viewed this for a while, come back to a contemplative state and think about those drops. What do you need in your life to be sure that you are well cared for? How can you nourish and nurture yourself? Take note of these things and pick one, two, or three of them. Finish your bath by contemplating how you can achieve these things in the mundane realm. When you are done, sprinkle the rose petals on your outdoor shrine. Then head back to your temple and write down the things you picked to change so that you can place them on your shrine. You can even place them in an actual chalice. Then commit to making these changes happen in the mundane world.

Exercise: Waning Moon Drowning Ceremony

The waning moon is a good time to release feelings and unwanted energies. While some use burning rituals to do this, as a water priestess, you can perform a drowning ceremony that works with water to accomplish the same thing. This is useful if you can't have an open flame or smoke in your living space. I suggest doing regular drowning ceremonies during the waning moon, but doing them anytime is useful.

Like burning ceremonies, drowning ceremonies are banishing rituals that release negative feelings, energies, habits, blocks, and thoughts. They can also be used in shadow work to help you shed and release things that hinder your own growth and progress. Like the Mermaid of the Dark, we are not always full of love and light, and sometimes these dark aspects of ourselves need to be brought to the surface, examined, and then drowned. Sometimes focusing on the energy of the depths of the ocean can give us perspective.

In this ritual, you will write what you want to banish on paper and drown it in a bowl of water till the paper disintegrates.

Then you give the water, the paper, and the energy back to the earth to recycle. If you work with mer spirits, you can also call on them to help banish and drown your sorrows.

For this exercise, you will need:

- A sacred vessel (I recommend a bowl that is dedicated to the waning moon or shadow work)

- Pen and paper

- Waning moon water made by charging water with the energies of the waning moon, much like Full Moon water

Place your sacred vessel in your work space. You may want to have a different vessel for this type of work to keep the energies different from those of your Full Moon healing and manifesting work. Pour the waning moon water into the bowl and charge it with intent by speaking to it and telling it what you need.

Write what you want to banish on the paper and charge it as well, then focus your intent and drown the paper in the water, crumpling it up, tearing it apart, and letting it turn to mush. This may take a while, so have patience. You can also leave it overnight for three days and then complete the ritual.

Take your vessel outside and pour its contents onto the ground. Bury the paper mush. If the ritual addressed personal work, emotions, or healing your local land spirits, let the earth recycle the energy. If you are working banishing magic, be very sure to dispose of the water and paper mush at a crossroads or in a location away from your home—in the west (endings) or north (underworld and death). Walk away, don't look back, and let the magic work.

Exercise: Waning Moon Cord-Cutting Ritual

This bath removes the energetic ties that you have to people, places, things, and events. These are usually focused around relationships, but cords can attach to anything. When seen through the mind's eye, cords that are golden, silver, or rainbow-colored hold higher energies and are usually considered to be good cords that shouldn't be severed. Some cords may be neutral; others may be dark and sticky like tar. These are usually connected to toxic relationships, traumatic past events, and other energies that may hinder your own growth. This bath can help you break those darker cords.

You can set this bath any way you like, just be sure that it is comfortable and that you can sink down for a meditation. You will also want to have room to move your hands to cut cords as described below. You can use a quartz wand, a smokey quartz wand or point, or a sharp shell to do this. Megalodon teeth also work surprisingly well!

Once you are in the bath, sink down and focus your intent. Journey to your astral temple through meditation (see chapter 4), then move yourself in your mind's eye to the sacred bath that is located there. It may be similar to the one you are in now, or it may take the form of your fantasy bath, an old temple, or a sacred spring. Once there, ask your spirit guide to show you where the cords are located, which ones serve you, and which need to be removed. When you have identified the negative or baneful cords, use the same tool you have in the mundane world to cut them and remove them from your body in the astral realm.

When you are finished, ask your guide for any other wisdom from the water realms and journey back the way you came. Once you are back in the mundane world, sit up in the bath and cut the cords in the same place as you cut them in the astral realm, then visualize a pure sacred light healing up the wounds and closing them off so the cords can't re-grow. Be sure to do this visualization often, as cords do sometimes re-grow. When you are done, stand up, step out, dry off, and drain the water.

Exercise: Waxing Moon Ritual Bath

This simple ritual helps you focus on growing a desire and bringing it to manifestation. Starting on the New Moon when you wake up in the morning, focus on your goal. When you drink your first glass of water, leave one sip and pour it into a sacred vessel on your altar space. Continue this for nine consecutive days. On the ninth day, charge your first drink of water and pour it into the same vessel. That same day, take a ritual bath, adding naturally colored orange rose petals and decorating it with items that represent your manifestation desires. Fill the tub with water and pour in the manifestation water you have collected over the nine days. Climb in, focus on your desires, and meditate on what you want to manifest. Focus your intent and see it manifest in your mind. Then sit back, relax, and enjoy your bath. When you are finished, plant the rose petals at the base of a tree or a plant you are growing.

Exercise: Waning Moon Ritual Bath for Releasing

You can do this ritual any time during a waning moon–but the closer to the Dark Moon, the better. I personally like to use black volcanic bath salts (do not use salts made with dye or ash). You may want to use a black bath bomb or other dark-colored bath additive. Set the tub with smoky quartz and black candles. Be sure to bring a paper and pen with you, and at least one black candle.

Prepare your dark bath, light the black candle, and step in. Spend time in contemplation, letting your mind wander to what you want to release. Focus your intent on it and see it dissolving in your mind's eye. Invite grace in and allow yourself to explore the feelings of release.

When you are finished, allow everything to fade away. Stand up and step out of the tub, but do not drain it yet. Dry off and write down any feelings, emotions, and words that correspond with what you have released. Light the paper using the black candle, let it burn in a dish that can withstand flame, and dump the ashes into the bath water. Once the ashes have touched the water, pull the plug and let the ashes disperse in the water as they swirl down the drain.

Exercise: New Moon Ritual to Nurture Growth

The New Moon is the perfect time to start work that has to do with growth and things you want to see come to fruition. It is a time replete with birth energy and can be a great time to work on bringing new energy into your life. This is a really fun bath, so get ready to nurture your growth!

First, gather seeds that you know grow locally in your environment. Later, you may want to work with a specific type of seed that is harder to grow, but for now, use seeds that you know will grow well in a garden or a small window box. Calendula, sunflowers, veggies, or daisy seeds all work well in most places. You may also want to work with herbs like sage, lavender, rosemary, or thyme. No matter what you choose, be sure that the seeds are safe to bathe with, and that you have a place like a window box, an outdoor shrine, or a garden in which to plant them.

Be sure that the bath water isn't too hot and that your seeds are organic. You don't want to bathe in water contaminated by seeds that are covered in chemicals. Organic seeds are quite easy to acquire. You will need a towel to dry yourself, but also an extra towel or sheet on which to put your seeds. Begin with plain water–no soaps, salts, or additives this time. You can add fresh flower petals if you like.

Place your seeds in a small sacred vessel you can take into the bath with you. You can even put them in a shell. Step into the tub and relax. Hold the bowl or shell of seeds in your hands and imagine yourself as a priestess, with all blocks removed and all obstacles overcome. See yourself as your best self, doing the work you want to do in the way you want to do it. When the water has cooled down a little, let the seeds float into it. Let them touch your skin. Let your mind wander to your priestess self again; see yourself blooming like a beautiful flower, producing seeds that

nourish others and heal the waters and any other aspect of yourself you want to nourish, nurture, and grow. Relax, and enjoy
your bath.

When you are ready, stand up, but do not dry off. Rather
step out of the tub onto the extra sheet or towel. Stand over it
and wipe, shake, and scrape the seeds off your body. Towel-off
standing on the extra sheet as well. Then gather a bit of the bath
water in your shell or vessel, drain the tub, and get dressed. Take
the extra sheet out into the garden or wherever you are going to
plant the seeds, and shake them directly into the soil where they
will grow. Cover them with dirt and sprinkle the water from your
bath over them for their first watering. Hold your hands over the
seeds and charge them with intent one more time. Bring blessed
water to your seeds every day and water them much as you would
nurture the things in your life and your priestess work that you
want to manifest and grow.

Exercise: *Charging Sun Water*

Charge Sun water on a Sunday by placing pure or sacred water
in a glass or crystal bowl and setting it out in direct sunlight for
nine consecutive hours. You can add plants, crystals, or symbols
with solar associations to enhance its Sun energies. When the
water is fully charged, bottle it and save it for your solar rituals
and workings.

Exercise: *Sacred Sun Water Bath*

The Moon may be a powerful force in your life as a priestess. But so is the Sun. And just as Moon water can play an

important role in your rituals, so too can Sun water. Sun water is simply water that is created by charging it with solar energies.

For this bath, charge your water for at least three hours in full sunlight. You can use this water for a variety of practices. I personally like to add sunstone to my water as it charges. However, do be sure to use the indirect method if you decide to ingest water infused with sunstones (see chapter 4).

Exercise: Sulis Sun Salutation

Each morning, take a vessel of spring water and place it in the sunlight. Let it sit for a few minutes with the light flowing directly on both you and the sacred vessel. Take nine sips of the water, then sit and meditate on the light and power of the Sun. Call on the Celtic goddess Sulis, who is the goddess of both thermal springs and the Sun. Meditate on the power of solar water and the light that refracts and glitters on its surface. When you are finished, take nine more sips, then gift the rest of the water to the earth where you are doing your ritual.

Seasonal Celebrations

Well-dressing is a very old folk practice whose roots probably go back to ancient times. Wells were dressed in recognition of particular seasons or holy days. In the UK, many sacred, holy, and magical wells and springs are decorated according to the calendar of sacred sabbats. The Chalice Well at Glastonbury, for instance, is decorated to celebrate Beltane, Lughnasa, Samhain, Yule, and Imbolc. Wells that were once Pagan, but have been adopted by

Christian churches, are often decorated on sacred days that align with the saints, as on St. John's Day.

If you live near a sacred well that allows you to engage in these practices, you are very lucky; I am sure that working to honor this place will become an important addition to your magical work. But if you live near a well where local laws prohibit this type of decoration—or if you don't live near a well at all, but still want to engage in these practices—you can create your own sacred well.

To do this, use a small circular table and a sacred vessel. You can change the sacred vessels out depending on the season, using black for Samhain, green for Yule, silver or white for Imbolc, red for Beltane, etc. No matter what you choose, be sure that the vessel is round and fairly deep so it can hold water for the entire season, or for a few days during the seasonal celebration. I have found deep meaning in this practice. The ability to honor the water spirits in season, while charging and creating seasonal waters to use in ritual, scrying, seasonal blessings, cleansings, or healings, has become an important part of my practice.

Here is a list of some of the seasonal festivals and holidays you may want to celebrate:

- *Samhain*: For this celebration, which marks the end of the harvest season, choose a darker-colored bowl—perhaps black, dark blue, or even a deep red—then decorate with sprigs of rose hips, hawthorn, apples, or other plants associated with this time of year. You can also add smoky quartz, garnet, and other crystals that are darker in color to connect with the nocturnal energies.

- **Winter solstice:** Also known as Midwinter, this is the time when days are short and the Sun hangs low in the sky. It is a time to pull on the celestial energies, so choose a silver or gold bowl or perhaps tie in with holiday colors like red or green. You can decorate with pinecones, evergreen branches, poinsettias, and strung cranberries, then add quartz crystals, and red and green candles or LED lights.

- **Imbolc:** Imbolc marks the coming of spring. This beautiful festival connects to the Celtic goddess Brighid, or St. Brigit, who is associated with sacred springs. You can decorate your well with a white cloth, a golden bowl, and candles to honor her. Add a few sprigs of fresh rosemary, a few snowdrop flowers, and early-blooming daffodils to complete the dressing.

- **Spring equinox:** The spring (or vernal) equinox marks a time of celestial balance and the passing of winter, so make this your most colorful well-dressing! You can decorate with fresh flowers, beautiful spring leaves, or newly blossomed roses, then add painted wooden or ceramic eggs. If you have a lot of patience, you can use real egg shells. Just poke small holes in each end, blow the yokes out, and paint them. You can keep the eggs on your shrine as vessels of your intent after the season has passed. Or you can turn this into a beautiful way to manifest things by placing your wishes with a few seeds inside the empty egg shells. At the end of your celebration, plant the eggs, crushing the shells and exposing the seeds so they grow into flowers with your wishes wrapped around their roots. Decorate the eggs with natural and food dyes so that you don't introduce glitter and acrylic paint into the earth.

- *Beltane:* This festival, traditionally known as May Day, honors life and the beginning of summer. Choose a red cloth and a complimentary bowl for dressing your well, then add hawthorn, which is usually in bloom at this time, and other freshly blossomed flowers that align with these energies. Beltane is a Fire festival, so you may want to decorate with white or red candles. Dew that is collected just before dawn on May Day (May 1) is considered sacred and was used in healing rites as well as beauty rituals in British folk magic, so add a few drops of dew to your sacred well water.

- *Summer solstice:* Also known as Midsummer, this is a time of warmth that connects us to the Sun. It marks the point at which the Sun is at its height. I personally associate this day with the goddess Sulis, so I decorate in golds, sea green, and teal, which is similar to the color of the waters in her temple at Bath. However, many also associate this celebration with St. John and harvest St. John's wort, with its little yellow flowers, on this day. Do what feels right for you.

- *Lughnasa/Lammas:* Also known as Loaf Mass Day, this celebrates the first grain of the harvest. It is a time of gathering in and giving thanks. In Celtic folklore, this day is associated with the Welsh Lady of the Lake—not the lady with whom you are most familiar in the Arthurian romances, but rather the heroine of a story that dates from a much earlier time. Decorate the Lammas well with grains, corn dollies, herbs, and the flowers of late summer. You can also add small cookies, oat cakes, and wheat weavings.

- *Fall equinox:* This is another time of celestial balance, one that reflects bounty and the beginning of autumn as the days begin to grow shorter. Decorate your well by placing early golden and red leaves around your sacred vessel and add plants you have harvested from your garden or your favorite fall foods. Gourds, small pumpkins, and other harvest vegetables make good additions, as do crystals like citrine, orange calcite, jasper, and carnelian. Wheat weavings, small colorful corns, and sunflowers can also energize the display.

Exercise: Creating a Sacred Well

Cover a small round table with a cloth that aligns with the season or energy and place a bowl in the direct center to act as your sacred vessel. Then choose your water (I almost always use water from a spring) and pour it in, charging the water with your particular intent. If you want to do a ritual on a particular holy day, or in a particular month, or during a particular phase of the Moon, place your intent then. If your intent is a complicated one, write it on paper and place it under the bowl for the entire season.

You can use small LED lights or tealights to illuminate the well and set the mood. LED lights can bring in a particular color that aligns with the season, and they are safer than tealights if you decide to keep them burning for an extended period of time or while you are away from your altar. I personally like the battery-powered strings of lights that come in different colors because they are easy to wrap around greenery.

Next comes the fun part—decorating the well! Start with bigger foliage—evergreens for winter, golden leaves for the fall, yellow flowers for the summer, grains for the harvest, flower blossoms for spring, and Brighid's cross for Imbolc. Focus on flowers, greenery, and fruits that are in season. Then add spiral shells, ammonites, buttons, coins, and other appropriate symbolic items used in folk practices. You can also use corn dollies, wheat weaving, spirals, and other objects with well correspondences.

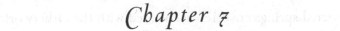

Chapter 7

WATER DIVINATION

*A*cross many cultures and throughout history, water has been known to have a liminal quality. Water in esoteric traditions is often associated with the Western Gate, the portal into the Otherworld and the world of the ancestors. It was also well known to be the home of many spirits, nymphs, mermaids, water goddesses, well guardians, and ancestral spirits. Where the water meets the shore has long been viewed as a liminal crossing point, as is crossing over water using a bridge or boat.

In Celtic lore, there are stories that describe magical women called the Gwragedd Annwyn who lived in lakes. One later legend with Otherworldly overtones tells of the Witch of Lok Island, a woman with ties to the faery world who lives on an island that can only be accessed by a swan boat that dives deep under the waters of the lake. In fact, water has been used as a portal in many folk traditions, while mirrors, like the smooth still surface of a lake, are often associated with the merfolk and scrying messages from the Otherworld.

Sacred springs are often associated with the Otherworld as well, and with divination in particular. St. Madron's Well in Cornwall is one such spring where dreamwork is done near the well. We know today that the well puts off radon gas that helps induce sleep and prophetic dreams. The branches from trees associated with water, like willow and hazel, are often used as divining rods to search out where a well may be dug or where a spring may be present underground. Springs and sacred wells contain high vibrational energies that are picked up by the rods, which were shaped like the letter "Y." They were held by the top prongs while the bottom leg indicated the presence of water through movements or dips. Similar rods were also used to divine ley lines and energy portals. Today, we use metal rods in place of the old folk methods. In this case, the rods are held lightly in both hands, indicating energy lines when they flick inward and cross.

Most types of divination involve trusting your own mind to correctly interpret what you hear, see, or sense. Trusting your visions, your intuition, and the voices you may hear in your mind is key to deepening, not only your divination skills, but also your connection to spirit. Not all that you see and hear will be accurate. Most card readers are happy with an 85 percent or better success rate, because they know that their visions depict only one possible future. Visions are not always certain, nor is their interpretation. In my own work, I have found that, although my visions are always clear, I may not always interpret them correctly. This is a skill that must be learned.

We all have the power to activate our own psychic abilities. Many were born with them already active and have vivid childhood memories of hearing voices, seeing spirits, or other supernatural phenomenon. It is societal conditioning that makes us doubt the reality of these experiences. Many of us are shamed into slamming the door on these abilities never to open it again.

As priestesses, it is vital that we honor the water and spirits we work with. But we must also become trusted conduits for their energies by living in sacred union with them and tending their temples—our own physical temples, the temple of our bodies, and the natural temples where the water flows. When in these states of sacred union, we become a vessel through which the spirits can speak. We are responsible for relaying messages from the water and water spirits to others in order to help them heal, or to take actions that will ultimately help them and thus the water. When we work consciously with the sacredness of water, we bring harmony and healing back to the world. And when people return to harmony with nature, the sacredness of water will be restored. We must, of course, heal ourselves before we can help others find this balance.

Exercise: Tending Your Sacred Temple

Perform this ritual nine days in a row to cleanse your own energies and prepare yourself to divine with the waters.

For this exercise, you will need:

- A sacred vessel

- Something to cover your vessel

- A chalice or cup

- A photo of yourself

- Drinkable water

Place your photo on your altar and put the empty sacred vessel on top of it. Each day, fill the chalice or cup with drinkable water and speak to it, using your voice and speech to charge the water with the quality of whatever you want to give yourself that day—peace, hope, rest, love, etc. Speak into the chalice nine times, then pour a splash of the water from the chalice into the sacred vessel. Cover the vessel to keep the water in it clean, then drink the remainder of the water in the chalice.

Repeat this each day for nine days until the spell is complete. Give yourself nourishing words on each day of the spell. On the final day, uncover your sacred vessel and drink the water you have been pouring into it each day. Feel free to modify this for your own use.

Trust Your Mind

Our brains are made of tiny cells called neurons. The neurons communicate with each other using electric pulses. When these cells communicate, they send multiple signals to the brain. These electrical currents are then regulated by the brain in the form of waves, creating what is called a brain-wave pattern. These waves are registered in the brain and emit tiny electrochemical impulses of different frequencies, creating different mental states known as Beta (12–38 Hz), Alpha (8–12 Hz), Theta (3–8 Hz), Delta (.5–3Hz), and Gamma (40+ Hz).

Beta waves support our normal conscious state and how we function in the mundane world. In Beta, we are relaxed and focused, and can process positive thoughts easily. In Alpha state, we are relaxed and can achieve a light altered state. This is where psychic work, meditation, imagination, and most scrying take place. In Theta state, we can achieve deep trance and deep relax-ation, which can support astral travel, shamanic trance, light ritual possession, and lucid dreaming. Delta state supports very deep sleep, very deep trance, ritual possession (without remem-brance), and coma. We can have trouble remembering what takes place here. In Gamma state, we experience ecstasy–divine inspi-ration and creative genius.

You will pass through and utilize all of these states as you walk the way of the water priestess. The water and its spirits will speak to you, appear to you, touch you, teach you, and draw you in. The key is to learn to trust your visions and your intuition. It doesn't matter what type of divination you choose to practice, or even how you execute it. The important part is that you trust your mind, develop your intuition, and hone your skills so that they become part of your craft. The water will do the rest.

In this chapter, you will learn how to work with a variety of divination practices–from hydromancy to casting bones, from scrying to channeling, and, yes, even a little ritual possession. What binds all of these together is that they are all connected to water in one way or another. And they all require that you learn to trust your own mind and honor your intuition.

Exercise: Channeling Water's Message

This practice begins with you sitting in front of your shrine or altar. Be sure to sit comfortably and support your body in any way it may need. Pain and discomfort can inhibit your success in this practice. Enter into meditation and take yourself to your astral water temple. Commune directly with the water or water goddess, then bring the images and messages you receive back and share them with others however you wish. You may also experience messages in a light trance state without visiting the astral water temple. They can come in images, sounds, and feeling. Sometimes you will speak with accuracy but not remember the message or process, which may mean you have moved into ritual possession (see pages 180–182)

I recommend doing this as part of your regular work and daily devotional practices. If you are dedicated to a particular water deity, it is best to start there and begin to connect through the meditation. I also recommend reading books on how to develop these skills or taking a workshop at your local metaphysical shop.

Exercise: Prophetic Dream Pillow

Another simple way to divine water's message is to use a pillow that is specially constructed and blessed to encourage messages to come to you through your dreams.

For this exercise, you will need:

- Dark-blue fabric cut into 2 pieces either 3 or 6 inches wide
- Mugwort
- Wormwood
- Eyebright
- Vervain
- Amethyst
- Moonstone
- Herkimer diamond

Paint, draw, or sew an open eye on the fabric pieces and sew them together with the images facing inward, leaving one side open. Turn the pouch inside out so that the images appear on the outside and fill it with mugwort, wormwood, eyebright, and vervain. Add the amethyst, moonstone, and Herkimer diamond, making sure that the stones are in the center of the pillow to keep it soft. Stitch up the fourth side of the pouch and bless it with the sacred waters of your choice.

Moldavite is also known to help with dreamwork and opening the Third Eye. However, this is a very powerful stone to which many people have strong reactions. If you are used to working with moldavite, consider incorporating it into your pillow. If you are not familiar with it, work with it first during the waking state to see if you will be comfortable working with it during dreamtime.

Hydromancy

Hydromancy, simply put, is divination using water by interpreting the movements on its surface when it is disturbed in a ritual manner. Water divination is an incredibly old practice that is still practiced all over the world today. One form we have already mentioned, called water witching, was used to find water on a dry land surface by interpreting subtle vibrations in a divining rod. Other versions of this practice used two different sticks that crossed over each other when water was present.

In the British Museum, there are two "Druid spoons" that may have been used for this purpose. Some anthropologists and archeologists believe that these spoons were used in divination, though there is little evidence to support the theory at this time. Yet the spoons were clearly not used for eating or for consuming anything liquid, as one of them has a single hole in the center, and the other is split into four sections by a decorative cross-like motif. The spoons are made of bronze and do not have long handles. Researchers believe that water or blood was dripped from one spoon onto the other through the hole. Then someone skilled in divination interpreted the meaning based on the section of the lower spoon that caught the liquid drops and what shape they formed there. Usually, these types of spoons are found in pairs. The ones in the British Museum were found in a bog near a spring.

There is another particularly interesting form of folk hydromancy from the Scottish Highlands that uses an egg, a pin, and a glass of water. On Halloween, or Samhain, an old sybil performed a ritual in which participants pierced a raw egg with a pin, then held the egg over a glass of water, letting a few drops of raw egg yolk fall into the water. The sybil then interpreted the meaning based on the shape of the egg yolk.

Scrying

Scrying is the art of gazing at a surface and deriving meanings from the images or shapes that appear there or in the mind's eye. Since this is a book on water magic, we will focus on scrying with water, but you can follow these instructions if you decide to try any other form of scrying as well.

Scrying is usually achieved through induced trance in an Alpha state. You can do this with a still body of water, a bowl or cup of water, or even your bath water. You can use regular water, or create a custom blend, or use water from a sacred well, a river, or the ocean. You can scry with just the water or put mirrors in the bottom of a vessel. You can add ink, crystals, wax, or herbs to the water, and even experiment with certain potions. Some use a black surface or a black mirror, while others use a crystal ball with the intent of telling the future. Some use a blue bowl to connect with water; some use a silver or crystal bowl. You can even use a still well or the calm surface of a lake.

Scryers usually sit in a comfortable position gazing at a shiny surface until images appear in their mind's eye or on the scrying surface. Once the trance is induced, they may see, hear, or feel images or sensations that they then interpret. Symbols, animals, pictures, plants, and other images take on special meaning and significance, depending on the intent. Scrying can be very taxing, but I have found that it gets easier with practice. It requires that you let go of your mind and allow images to come through. These can be difficult to interpret, but, with practice, you will learn to trust your intuition. Most Priestesses have a favorite scrying tool–a bowl, a black mirror, the surface of a lake, or a crystal ball. In fact, any reflective surface will work.

Before you begin any scrying activity, be sure to make yourself comfortable. It can be hard to achieve an Alpha state if you are distracted by bodily aches and pains. Sit on a pillow or chair with your bowl or other tool placed comfortably in front of you so that your neck is not strained. Begin by breathing in a manner that will help you relax. I like to take nine deep breaths to connect with the energies of the water priestess and Delphi and the Muses. Select your sacred scrying vessel and your sacred water depending on your intent. You will also need your Book of Tides, a few scraps of paper, and a pen or pencil.

The scrap paper is really important, because you don't want to scry and write in your Book of Tides at the same time. But you still need to record the information that is given to you without breaking your gaze or state of mind. Just place the scraps of paper beside your scrying vessel near your dominant hand. You may want to tape them down so that you are not distracted by moving paper and so that your hand can freely move over the paper and write the words, symbols, images, phrases, feelings, and visions you may receive during the session. Once you are finished, you can transfer your notes into your Book of Tides, where you should keep a very detailed record of all your scrying sessions. This will help you find your strengths as well as your weaknesses, and help you know what will and will not work for you moving forward. There is no reason to reinvent the wheel; take the path of least resistance and flow where the water takes you. You may be surprised at what comes easily for you and what doesn't.

When you record in your Book of Tides, be sure to put the date, time, Moon phase, planetary hour, how you feel, what type of water you used, what type of vessel you used, and how the session went for you. Over time, you should be able to establish a pattern by looking through the book and finding what has worked. This will help you to move forward and begin to feel more confident in your practice. It will also help you track which prophecies come true.

Scrying is probably one of the hardest forms of divination, because it relies entirely on your intuition and the shapes that appear in your mind's eye. After time, the simple shapes may become more complex—like movie clips or trailers—and that makes interpretation much easier. If you have tried scrying and found it too frustrating, try some of the fun suggestions below. Sometimes adding ingredients is all you need for a bit of success!

Exercise: Creating Scrying Water

You can create your own special water to use in your scrying practice. The properties of the ingredients you use can help to focus your intent and bring forth clearer and more powerful messages.

To create this water, you will need:

- Eyebright

- Hawthorn

- Herkimer diamond

- Marigold

- Mugwort

- Rose

- Spring or ocean water

- Thyme

- Vetivert

- Wormwood

Infuse the spring or ocean water with some or all of the herbs listed and then add the stone. Let it sit for three to nine hours, then remove the stone, strain the water, and use right away. If you wish, you can preserve the water for altar use with vodka or brandy.

Exercise: Scrying with Water

When you are ready to begin, place your scrying vessel in front of you, light candles, and turn off the lights. Then pour your scrying water into the vessel, leaving room at the top. Go through one of the grounding exercises given in this book, then find a comfortable position and settle in to begin your practice.

Start by clearing your mind and relaxing. Don't try to rush this, and don't get too disappointed if you don't get any images at first. It took me a good two years of practice before I became skilled at scrying. When you are relaxed, call upon any spirits or ancestors you may work with when doing oracular work. Ask them for clear messages and the wisdom to understand their meaning. Establish a rhythmic breathing pattern and allow yourself to slip into an Alpha state and let your mind wander. It may help to start this portion with your eyes closed so you can relax and clear your mind. Be sure that there are no mundane thoughts running through your mind.

When you begin to see images, jot them down on the scrap paper so you can record them in your Book of Tides later. Use these notes to help you eliminate messages from your ego and concentrate on true messages from the Otherworld. You will soon learn which is which and will be able to focus on the authentic messages rather than on what your ego may be telling you. When you are finished, thank your spirits and ancestors for their messages. Connect firmly with the earth and bring your mind back to the present moment. Record your notes in your Book of Tides and add thoughts, feelings, colors, and anything else you may have felt during the session. Over time, you will start to see patterns emerge and will become more skilled.

Exercise: Scrying with Ink

Inks can be a really fun way to begin to flex your scrying muscles and learn how to become skilled at reading the messages of the waters. I actually started experimenting with inks because I am an artist and wanted to know what would happen if I added a few drops of my inks to the water. I began by adding a few drops of black ink and letting it naturally move through the water. I used it to help induce a trance and to bring forth particular shapes. This really helped build my confidence and made it easier for me to move to a more traditional water-gazing practice. I am much better at scrying than I was back then, but I still love playing with different colored inks. My personal favorites are light blue in clear water and silver in a black vessel. You may need to give the water a spin if the ink doesn't disperse easily. I use a little piece of wood for this and just give the water a push. Then I sit back and watch the ink take shape. In the past, scryers dropped molten lead into the water and interpreted the shapes that it formed as it solidified. But ink is safer and easier to use.

You can also use magically infused oils in this way. They can be particularly effective and can bring added properties to the session. If you want to make a natural ink, try crushing blackberries into a thick juice and using small drops. Or you can boil herbs to produce darker-colored water that makes a nice contrast with the clear water. Fresh flowers floating on the water's surface or dried herbs sprinkled there can also bring forth messages. Try sprinkling a bit of mugwort or eyebright onto the surface and see what happens!

Exercise: Scrying with Water and Fire

Water scrying can be tricky! Sometimes adding a candle close to the bowl so that the flame dances across the water's surface creates a useful play of light and shadow. The candlelight creates reflections that you can use to interpret the water's messages. Choose the type of water depending on your question and intent. Consider using Moon waters, Sun waters, or gem elixirs. You can also drip melted wax from a candle into the water and inter-pret the shapes that appear.

Exercise: Scrying with Balls and Mirrors

Many priestesses use onyx, selenite, calcite, or clear quartz balls rather than vessels of water to scry. Others use glass in a variety of colors. Blackened mirrors or mirrors lit by candles can also make good scrying surfaces. As with water, these reflective surfaces are used to induce an Alpha state. Once in the trance state, diviners see images in their mind's eye and interpret the meanings.

To bring a watery aspect to this, you can add a stone or crystal to the center of a bowl of water. First be sure that the water won't harm or break down the stone. I recommend working with quartz or onyx, but you can experiment with other stones as well. Or try casting water over a mirror and looking at the way the drops form. You can even place a small mirror at the bottom of a bowl or use a silver bowl that is very reflective.

Exercise: Scrying with Sacred Water

For this exercise, you will need:

- Rose

- Hawthorn

- Mugwort

- Lake water

- A hag stone

- A quartz pebble

- 2 sacred vessels

Infuse the lake water with the rose, hawthorn, and mugwort, then add the hag stone and quartz pebble. Let the mixture sit for thirty-six hours. Add some vodka to preserve the water and keep it in a special bottle. When it is time to scry, pour the water from one of the sacred vessels to the other and consider the patterns it makes as it flows.

Exercise: Scrying on a Lake

Lakes have the unique property of stillness, which differs from most other forms of liquid water, which flow. Steam is constantly in motion, and even ice shifts and cracks and morphs as temperatures change. The stillness of a lake, on the other hand, provides a clear glassy surface like a mirror that can be used as a scrying tool.

Mirrors, especially black mirrors, are often used in magical practices. We also know from folklore, faery tales, mythology, and even pop culture that mirrors can act as a type of portal. A lake in a still state becomes a mirror of sorts and can provide a wonderful surface on which to scry. Local man-made or naturally occurring ponds also can be used in this manner. There are several different ways you can do this. If you find yourself at a lake and feel the desire to receive divinatory messages from the water spirits who dwell there, spread a blanket at the water's edge and begin to move yourself into a trance state. If you are able to situate yourself on a hill or rock to get some height over the water, this can give you a higher vantage point so you have more surface to contemplate.

Just as when you work with your scrying bowl, you may see images play out on the surface of the water. You may see shapes that trigger vision journeys, or you may see images directly in your mind's eye. One way in which lake scrying differs from scrying with your sacred vessel is that you may see reflections of the sky in the water and the light may play across it, creating a sonoluminescent light show that moves you into an Alpha state in which you can receive the water's messages.

Exercise: Scrying the Storm

Storms—snow, rain, and hail in particular—are active water energies that can be scryed for messages from the waters. The patterns they naturally make can contain specific messages for you to interpret. Spending time watching these types of energies can give you great insight into weather magic and wisdom about how water plays a part in both the mundane and magical aspects of the watery world. For this type of scrying, place yourself in a safe and comfortable environment where you can observe these weather patterns. Record what you see in your Book of Tides and consider what you think it means.

Exercise: Scrying Ice

Scrying ice is similar to scrying with a crystal ball. However, you will end up with a much different type of message—one carried in the crystalline structure of the water. When you gaze at a piece of ice, you focus on the patterns frozen within it. Perhaps you concentrate on all the white parts, or perhaps on the more clear parts. Or perhaps you gaze at both and let your mind wander over them to enter into an Otherworldly mindset or trance state.

The messages of ice are much different from those of water. They teach us of structure and strength, but also of rigidity and the coldness of a hardened heart. You can use a large spherical ice mold to create a sphere of ice into which you can gaze. You can use icicles that form naturally in the winter, or even regular ice cubes floating in water. If you are able to visit an ice cave or get to see large blocks of naturally formed ice, take the time to scry these monumental ice structures for the lessons they contain.

Casting the Bones

Using bones for divination–also known as osteomancy or "casting the bones"–is an ancient theme in oracular traditions across many cultures. Here you will learn how to collect your own set of divination bones and how to use them to receive prophetic and magical messages. British sea witches performed osteomancy by combing lines into the sand and casting their bones upon them. One diviner was known to cast her bones into an upside-down tambourine into which she had drawn lines. Bone oracles in China used the scapula of an ox or an inverted turtle shell to divine messages, carving the questions onto the surface and heating it till it cracked, then interpreting the cracks. In Africa, bone readers called *sangoma* used bones, seeds, shells, nuts, dice, beads, and other small objects, casting them from a basket onto a mat. The details of these practices varied practitioner to practitioner.

If the idea of using bones in your craft does not resonate with you, try using shells. They have many properties in common with bones. In fact, I recommend spending time near the water finding shells, bones, driftwood, stones, sea glass, and other useful bits. I began collecting my bones long before I even knew that I could use them for divination. Most of my items came from the water's edge—some from a cave pool in Japan, some from a cave in Cornwall, and some that washed up on the shores of the Atlantic Ocean. Some shells I have even had since I was a child. When I brought all these pieces together, I found that I had a full set of tools and was ready to divine. After months of contemplation and really thinking about how European sea witches may have cast, I began to call them my "sea bones." Although they are not all from the sea and are not all bones, it seemed to fit!

To start your own set of water-divination bones, spend time at your sacred water source looking along the banks for small stones, pebbles, bones, nuts, seeds, bits of driftwood, fossils, shells, and anything that is about the size of a quarter or smaller. For our purposes here, we will call these all "bones." Some sources will be very abundant in these objects, others may not. Don't get frustrated if you only find a few things. Anything more than two is a start; you can begin to work with them using simple yes/no questions until you are able to gather more items.

Some of these items may break over time; others may disappear. But then new ones will make their way into your set. Some bones will always be there; others may not. That is just the nature of the art. If one of yours breaks, assume it is no longer a working part of the set. If a seed randomly sprouts, plant it immediately. While I doubt that will happen, you never know! You may also be able to purchase some items. I have a small ammonite in my set along with a vial of gold flecks that I bought because they seemed appropriate.

Casting the bones takes a lot of confidence. It may seem as if the items in your set have no meaning, or that there is no in-depth meaning to each bone. But you will spend time giving each of them deep meanings that resonate for you and you will learn how to read those that cluster together. The meanings you discern for each item will grow and expand with each reading.

There are a few specific things you will need for casting the bones. First, make or buy a throwing mat or dish. You can fill a tray with sand and comb it as the sea witches did, or you can create a custom mat. You can use an antique silver tray, or a piece of leather or fabric, or a tambourine, or even a flat basket. Pick whatever works for you; there is no wrong choice.

You will also need two large shells about the size of half a fist to represent you and the person for whom you are divining. If you are reading for yourself, use both to represent you, focusing one specifically on your question. You will also need a large scallop shell from which to cast your bones. Place all the items in this shell and then cast them onto your chosen surface. Although you can start with just two bones, you should eventually have at least five, and twenty is probably a better goal. The more you have, the more detailed your readings can be.

You may want to split your throwing surface into sections. These can represent the past, present, and future, or perhaps the four elements. Or you can leave your surface blank and just read the relationships between the bones. You may also choose to assign similar meanings to specific bones to represent these energies in your reading. It is really up to you how you structure your divinations. The important thing is to decide on one method and stick to it. When you start changing it up, you create doubt and errors in your readings. There may be a time, of course, when you want to make a change for a good reason and commit yourself to this new method. Be consistent, and leave yourself a little bit of wiggle room. You will be calling on the spirit of the waters, the ancestors, and the properties of the individual items as they play across your throwing surface.

When you assign meanings to your bones, use a notebook dedicated only to this practice. This will help guide you and make it easier to organize your thoughts. On page 176 are some suggested meanings, but do not feel that you must stick to them. When you are assigning meanings, let spirit guide you. If you get stuck, look at the shape of the item and consider its magical meaning, its mundane associations, and its color to help you find a suitable message. You can also take into account the mood you were in when you found it and where you found it. For example, I once found a piece of coral that looked like wings; it took on the meaning of "guardian angel." I found a few small seeds that were perfect to represent children, who are like little seeds waiting to be born and to grow. Driftwood with a male shape may be masculine, while a hag stone may look like a vagina and be feminine. A feather may represent a bird guide, while a clam shell may represent a hermit or a hidden meaning. Always go with your gut; you know what it means! If you can't decide on a meaning, tuck the item away. I have several bones that currently have no meaning attached to them and are waiting for one to be discovered.

Here are some suggested meanings you can assign to your bones:

- Anger
- Children
- Evolution
- Female
- Forgiveness
- Healing
- Job
- Love
- Male
- Past, present, future
- Seclusion
- Spirit
- The four elements
- Time
- Travel
- Wealth

You can also use any other meanings that resonate for you. If after you have cast for a while, you find that you are in need of more–or specific–bones, seek them out and add them to your collection.

Exercise: Cleansing the Bones

In this ritual, you will cleanse the bones you have gathered. You can also use this whenever you find new items in the sea or at the water's edge. Most of us use a cauldron for this work, but if you don't have one, you can use a large pot or even a mini-cauldron. I recommend using fresh herbs. I like to use mugwort and worm-wood for second sight and divination, but any herb that you associate with divination will work. Adding a piece of snowflake obsidian can also help cleanse and empower the bones.

For this ritual, you will need:

- Cauldron or large pot

- Fresh or dried herbs (mugwort, wormwood, vervain, and yarrow work well)

- Heat source (fire, stove top, or even hot water from a kettle)

- Sacred water, or your choice of a water like rose water or a gem elixir

- Your bones

First, identify a space to work in–for instance, outside over a fire pit or on the stove–then cleanse it as you would for any working. Once you have collected your bones, put a few cups of water into the cauldron. As the water starts to heat up, add the fresh herbs and plants. (Dried herbs are a good substitute.) Let the plants simmer for about thirty minutes, then lower or extinguish the heat so that the water begins to cool. While the plant material is still in the cauldron, speak the charm below and place all your bones into the pot. If something breaks or becomes unusable during the ritual, assume you were not meant to use it.

Don't boil your items. You just want to make a tea-like liquid to cleanse them. If some items are not suited for heated water, consider sprinkling the water on them instead, then placing them in the cooled mixture for several minutes.

Now it is time to chant. Focus on your intent, or even begin to ask the spirits to bless your work and help you divine. Once you are satisfied, remove all the items and rinse them off with rose water or a gem elixir of your choice. Set them in a safe place to cool and dry. Make sure that they are completely dry before using them or storing them; you don't want them to become moldy. Your bones are now cleansed and you can begin to use them in divination.

Bone-Cleansing Chant

Bless these treasures from the sea,
These beautiful gifts bestowed on me.
Sea bones of death, now give new life;
They tell me of love, pain, success, and strife.
Sea bones of divination, now ye be;
So it is now, with the powers of lake, river, well, and sea.

Exercise: Divining with Bones

To prepare for a reading, first cleanse your bones as shown, then prepare your space and mat. Place your bones in the scallop shell and set your throwing surface between you and the person for whom you are reading. If you are reading for yourself, place it directly in front of you. Take the shell that represents you and hold it to your mouth, whisper your intent into it, and ask the ancestors (or other spirit) for guidance. Then place it back in front of you.

If you are reading for yourself, begin your casting now. If you are reading for others, take the shell that represents them, hold it to your mouth, and whisper your intent into it. Ask the question and request guidance from their spirit guides, then place the shell on the mat in front of them. Sometimes it is helpful to have them hold the shell and blow on it as well. (Be sure to cleanse this well between readings.)

Shake the scallop shell and toss the bones onto the divination surface. Sometimes I let them roll out with a little kick to get them onto the surface. Do whatever feels right for you; just be sure to be consistent. Once the bones are on the mat, begin to look for patterns and items that touch. Items that are touching are directly related; items that are not touching are less important, but can still matter, depending on where they are in relation to other items. When you are finished, thank the ancestral spirits. Cleanse your bones and put them away.

Ritual Possession

Ritual possession can occur in various ways and in varying degrees. Sometimes, when it happens unbidden, a trained professional or religious authority must be called in to exorcize the spirit. Some people are natural mediums with the power to lend their bodies to spirits in order to receive messages from the Otherworld. Others train and work hard to acquire these skills. Practitioners of various magical and religious paths perform rituals in which they are completely possessed by a spirit, sometimes even speaking with a different accent or adopting different behaviors or mannerisms. In these cases, they often have no recollection of the event. In the Umbanda religion, priests and priestesses dance to become possessed by spirit and then pass on messages to those in attendance. On the other hand, dance can also be used to exorcize malevolent spirits. In the Middle East, the Zār rituals cast out spirits using ceremonies comprised of dancing and trance that can go on for days.

In some cases, malevolent spirits can cause illness, disease, and an overall degeneration of the possessed person's body or spirit. In other cases, a person may welcome possession by a sacred or higher being, as in Wiccan rituals that draw down the spirit of the Moon or Christian rites that seek visitation by the Holy Spirit, as described in the book of Acts. All this can sound pretty scary—because it is! But while I won't describe in detail here how to achieve ritual possession, I would be negligent if I didn't address this art, as many people are naturally drawn to it. And if you are working hard on your divination skills and making really good contact, there is a possibility you may experience one of these states.

Ritual possession is scary because we have come to see all possession as a negative experience. However, it can also be a beautiful experience that brings you closer to a goddess or spirit with whom you have worked for a long time and with whom you have a trusting relationship. Trust is very important here and you must be sure to establish a working relationship with a goddess or spirit long before you lend them your body to speak. As we saw in the previous channeling exercise, you become the conduit, the mouthpiece, of the goddess and tell others what she has told you. You pass on her words; you never speak for her—never.

However, a goddess may speak through you, and this is one form of ritual possession. It occurs when a priestess enters a trance state and begins to speak with authority as the goddess or spirit herself. In Wicca, the high priestess performs a ritual on the Full moon that pulls the energy of the Moon down into her. When she does this, she opens up to the energy of the goddess and becomes one with her—a merging of divine spirit and sacred priestess. It is important that you keep your own body and your sacred vessel pure and in a vibrational state that is conducive to the deity or spirit with whom you are working.

A word of caution: Always work in a protected space, especially if you begin to suspect you are reaching a level approaching possession. There is no book in the world that can prepare you for these experiences, although there are some that can give you knowledge on the subject. If you find yourself lost in these states or overwhelmed by them, go to your nearest metaphysical or Pagan shop and ask them to recommend a good teacher who can work with you and help guide you through these unique experiences.

Chapter 8

THE ART OF RITUAL

*I*mages of priestesses holding sacred vessels and pouring sacred waters upon an altar are etched into our collective memory. They prompt a yearning, a desire to watch the water flow forth, to commune in sacredness with spirit. For some of us, it goes deeper, calling to a memory or memories of lives past when we lived as priestesses in our sacred temples enacting rituals and performing sacred acts to honor the water and the nymphs and deities that dwell there. The sacred presence of a water priestess was a necessity for those who traveled from near and far to be healed by the sacred waters or to worship them in grace and hope.

Today, we are beginning to revive and recreate the water rituals of antiquity–pulling from the historical and archeological record when we can, but relying mostly on the motions of our memories to enact the old rites in a modern world. Some rituals, like pouring water onto an altar stone, may need to be modified to accommodate the times–for instance, if you need to keep your home temple space dry to protect carpet or wooden floors. But generally speaking, the ancient water rites can be adjusted to fit within our modern lives.

The Vestal Virgins poured water upon the sacred fire, perhaps to control the heat or the flames. In Greece, salt water, both directly from the sea and in a prepared solution, was used in ritual cleansings of people, clothing, and objects. The oracle at Delphi bathed in her sacred spring before her oracular work, and water temples were commonplace throughout the Celtic, Roman, and Gallic worlds. It was in these sacred temples that rituals were enacted. While we may never know exactly what these rituals looked like or how they took place, we have enough evidence to recreate ritual practices that flow easily and enhance the power and spirit of the waters.

We know that water rituals generally included several elements and actions that facilitated the working and helped create sacred space in which to connect with the sacred waters. These included:

- Anointing the altar

- Anointing the body

- Cleansing and clearing the energy of the temple or room

- Creating and using sacred waters

- Creating sacred space or casting a circle

- Divination and prophecy

- Giving offerings

- Healing services

- Holding space for those who come to the temple

- Oracular work (receiving messages from the water and spirits)

- Parting the mist, bridging the worlds

- Pouring water

- Prayer

- Preparing and bearing sacred vessels

- Ritual bathing and baptisms

- Singing, chanting, and using sound in conjunction with water

- Water ceremonies

In this chapter, we will explore how to incorporate some of these elements and actions into your own water priestess practice.

Casting a Circle

Sacred space can be created effectively by constructing circles or spheres of energy to protect you in the astral realm as you do your spiritual work. This becomes the sacred ground upon which you work your magic—a place where sacred and holy spirits will come to work with you and appear to you. Some people choose to cast circles when they work. If you are working beside a lake or at a beach, you can find a quiet spot and draw a circle in the sand. Simply drawing a physical circle is not enough to protect you, however. Think of it, rather, as a limit, a boundary, a template for your astral circle. There are specific actions you must take when you create a sacred circle. First, you must focus your mind by visualizing a circle, while simultaneously focusing your energy. Finally, you must walk the physical space of the circle.

When we speak of casting a circle, we actually mean casting a protective bubble. So when you visualize and direct your energy, be sure to cast the circle as a three-dimensional sphere rather than as a flat ring on the ground. And make sure that the center of the sphere lies where your feet touch the ground so that the orb protects you above and below. The most important thing is to make sure that your visualization is complete and strong. Once the circle is complete, you can move around in your space without worry of astral nasties or other unwelcome visitors.

You can place magical objects like crystals, small bowls of salt water, or shells around the circle's border to act as anchors. When visualizing the circle, attach your energy to these objects to empower them. Have fun when choosing these and other physical objects to enhance the power of your visualization and anchor your astral circle. Try using drawings in the sand, stones, flowers, driftwood, salts, and quartz fragments. You can even use white sand from the drier portions of a beach to make a circle. Avoid wet sand, however, as it does not flow well.

Circles are very useful when working with other people and in large groups. When I am working with a group of people whose energy I cannot control, I use a salt circle. Circles are watery by nature–resembling the sacred vessel, cauldron, or bowl–and this allows them to bring in more watery energy. See chapters 4 and 9 for information on the physical creation of circles.)

Daily Ritual Practice

Daily rituals are meant to be short repetitive rites performed on a regular basis with a specific goal in mind. They may be devotions to a deity, a spirit, or an astral being, or ceremonies to charge your own sacred waters with a particular energy or to set an intention for the day. In your daily practice, you may choose to work directly with the water spirits, or with powerful deities associated with the water like Aphrodite, Sulis, Melusine, and Poseidon. Or you may consider working with ancestral priestesses, who make wonderful guides upon the priestess path. In my own experience, I have found them quite willing to guide and give advice. You can also choose to work daily with a particular body of water like the ocean or a local river, or with the universal spirit of water in a devotional practice. This will not only help you build your relationship with the water spirits, but will also deepen your connection to water in general and help you connect with it spiritually.

Daily rituals can be as simple as blessing a cup of water and pouring it into a sacred bowl on your water shrine each morning as an offering to a water deity, or saying a short prayer or blessing before drinking your first glass of water each day. Engaging in short and light meditations in your morning shower can also connect you with the water spirits. Or you can light a candle on your shrine and spend five minutes with your water goddess first thing in the morning, intoning a short prayer or sitting in meditation. Anointing yourself with water every morning–perhaps choosing a sacred symbol like a bind rune, a crescent Moon, an alchemical glyph, or a spiral to enhance the rite–can be a great way to start each day.

Keep a small bowl of blessed water (or protection water) near your bed at night and gift it to the earth in the morning. By keeping the water near your bed, you protect your dreams in the astral realms and, in the morning, you can offer it to the earth with gratitude, pouring its life force into the ground and nourishing the plants. You can also charge water with intent to water house plants using an oracle card from a watery-themed deck.

If you want these devotional practices to go deeper, perform them as full rituals every week or perhaps on every Full Moon. You may organize your daily rituals so that each day is dedicated to a specific intent, or you may want to perform them on specific days. Sunday may be the day you create Sun waters for the week; on Monday (Moon day) you may connect with lunar energies; Tuesday may be for prayer; Wednesday may be for healing. It is up to you, of course! Go with the flow of your own path and allow the waters to show you the structure that works best for you.

You can perform monthly rituals in your home, in your temple, in the bath, or at the water's edge. These can be as simple as dedicating the first Monday to water, or it may mean that you pick a particular ritual to do each month. For example, you may choose to perform purification rituals only on the Dark Moon or prosperity rituals only on the Full Moon. You may save balancing rituals for a quarter moon. You can give your rituals a more poignant tone by casting protection or healing toward a specific type of water or water creature like dolphins or whales. Taking a monthly sacred bath can also be a good way to honor your practice and to fill your cup. Ritual bathing can be a relaxing experience, as well as a sacred one. A lot of magic can take place during this watery act.

Blessing Rituals

Blessings come in many different forms and can be enacted in several ways. I have found that the best way to bless is to use several methods in conjunction. Here are some techniques, energies, and symbols commonly used in water blessings.

- *Anointing:* This action is most commonly described as the smearing of liquid onto an object. But it can also be done by sprinkling the liquid upon an object or person. It is usually accompanied by a blessing prayer or a prayer of consecration dedicating the object or person to a particular service. It may include drawing a sacred symbol on the item to be blessed. Objects that are commonly anointed include divination shells, bones or other divination tools, and healing crystals dedicated to water.

- *Asperging:* This is the act of sprinkling holy water. You can asperge altars, shrines, and temple surfaces, as well as the human body and its aura field. The water must be blessed and the herbs it contains must be holy and consecrated so that their spirits are awakened and attuned to working with the holy water.

- *Awen:* This is the spark of creative energy, the gift of divine knowledge. It is often used as a sacred symbol of inspiration and has been described as the flow of spirit.

- *Crescent:* This shape represents the crescent phase of the Moon, whose lunar energies are deeply tied to water. This symbol is often drawn on the forehead to represent the priestess path or dedication to the Moon or a particular Moon goddess.

- *Saining:* Saining is the act of blessing, protecting, or consecrating. Water is usually the main element used, however juniper smoke also works well. While this word is used to describe a specific cultural rite, the practice itself is universal, although you may find that it has a different name and its own particular cultural flair in different settings. Perhaps you will devise your own name for it!

- *Sound:* Sound and water are intimately connected, because water absorbs the energy of sound's vibrations. Because of this, prayer, chanting, spoken words, singing, and even instrumental music are wonderful ways to bless a person, place, or object when used in conjunction with water.

- *Spiral:* This shape mirrors the ammonite or nautilus, and is a symbol that can be used to represent longevity, opening, and cycles. If you want to close or seal energy, be sure to spiral from the outside inward; to open energy, spiral from the inside outward.

- *Trident:* The trident is a cross-cultural icon that has been used as both a symbol of power and as a weapon of protection. It is the weapon of choice of Poseidon and has been used in mudras and other protection marks. You can also use multiple tridents in a circle as an emblem of protection, as the symbol is incorporated into the Helm of Awe, also called *Ægishjálmur* in Icelandic and *Œgishjalmr* in old Icelandic.

- *Triskele:* This symbol has been used in many ways and can be found on flags, churches, stone carvings, and artifacts. It can represent movement, transformation, the three celtic worlds, and the three states of water.

- *Vesica Pisces:* Two circles of equal size that intersect create a shape in the center that is called the Vesica Pisces, which means "bladder of the fish." This symbol is used to represent fertility, sacredness, and femininity, and is often associated with the shape of a volva. It has also been used in illuminated manuscripts as a type of body aura called the "aureola."

- *Water glyph:* This symbol is used in alchemy to represent the element of Water. I use it most often to represent water or to activate the Water element.

Exercise: A Simple Blessing

This simple blessing can become a part of your daily ritual, or you can use it anytime to invoke the spirit of water. Gather the type of sacred water you want to use and choose a sacred symbol. Then select the object or person you intend to bless and decide where you will place the blessing. The blessing will encompass the entire body or object, but you still should find a central point or points on which to focus the work.

Prepare an incantation, prayer, song, poem, or other spoken words. Try to keep this short so you can repeat it; but long, deep heartfelt prayers work as well. If you work with a water goddess, call upon her now to lend you her healing powers and wisdom.

Dip your finger into the water and begin to draw the sacred symbol on the location you have chosen. Draw the symbol three or nine times—either on different sides of the target or overlaid in the same location—until it has been imbued with the symbol's energies. Thank the water and the energies that aided you.

Exercise: Nine Sacred Waters Ritual

This ritual connects you with nine different forms of water to create a very potent blend that you can use in your blessings. You can use the same water in the dedication ritual found in chapter 9.

This ritual is done over nine days and combines nine different types of water, including spring water, ocean water, river water, silvered water, holy water, sacred water, Full Moon water, Sun water, and eclipse water. You will need to collect all these waters in specific ways, so be sure to leave time to plan and prepare for this ritual if you want to perform it in a particular astrological cycle.

You will need a bottle or jar in which to combine the waters. If you are working with a small vial, use nine drops of each water. If you are working with a larger jar, use three splashes of each water. I suggest making larger amounts, since the preparation is time-consuming and you can save the water to use later.

Gather all your waters and line them up in the order given on page 194 in front of your sacred vessel on your shrine or altar. Begin by burning a sacred wood or incense. Myrrh is a good choice, as it aligns with water. Bless and dedicate the waters and cleanse any negative energy they may contain. Asperge the waters and the vessel you are filling with sacred smoke from the incense. You can use drumming, singing bowls, chanting, and other sounds in place of smoke if you prefer.

Once the waters are blessed and cleansed, pour the water aligned with Day 1 into your vessel, then use it to anoint your Third Eye with a crescent shape and your heart center with the alchemical glyph for water. If the water is drinkable, you can take a sip; if not, skip this step and be sure to label both this bottle and your final bottle as undrinkable. Do this each day for nine consecutive days, following the order given below.

- Day 1: Silvered water

- Day 2: Spring water gathered from your favorite spring

- Day 3: Full Moon water

- Day 4: Sacred water created by adding three pinches of salt to your choice of water and twirling it sunwise three times

- Day 5: Ocean water collected from your favorite beach

- Day 6: Sun water

- Day 7: River water collected with the current from your favorite river

- Day 8: Holy water collected from a sacred well, church, shrine, or sanctuary; or create holy water with spring water

- Day 9: Eclipse water set out in a bowl during an eclipse

On the final day, seal the bottle that contains your combined waters and de-construct the ritual. You can use this water in any appropriate ritual or working. Consider making it on your birthday each year to align it with your own energies.

Exercise: Gratitude Ritual

For nine consecutive weeks on the same day of the week, write the things you are grateful for on a piece of paper, each time lightly blessing the paper with sacred water. On the final day, write your list, but don't anoint the paper with water. Instead, carve the word "gratitude" on one side of a blue candle; then carve the alchemical glyph for water on the other. Anoint the candle with sacred water and light it. Then burn all the paper. Be sure to catch all the ash in a small bowl so you can release it with words of praise and gratitude on a windy day.

Sacred Bathing

Sacred bathing can be done just about any time to ground and purify your own energy—daily, weekly, monthly, on sacred days, before rituals, or after working with clients or hosting a Full Moon water circle at your local community center. No matter why or where or how you perform them, sacred baths can benefit the body, the mind, and the soul.

Sacred baths can be used for a variety of purposes and they can be done in many ways. You don't even have to perform them in a bathtub. Immersion in other bodies of water works as well—dipping in the sea, river rebirth ceremonies, and baptism are also types of ritual bathing. Sacred bathing can even be done completely clothed outdoors!

Cleansing and purification are probably the most common sacred bathing rituals, although protection baths, spiritual healing, love drawing, prosperity, dedication, curse and spirit removal, and other magical intents can also be empowered through them.

Exercise: Rebirth Bath

This bath connects you with the energies of rebirth. Many on the spiritual path, especially many priestesses, go through a process of death and rebirth that can be seen as part of the ending stages of a spirit initiation. This process involves the person being completely stripped of ego and going through an egoic death in order to be reborn to greater purpose. This often includes physical changes in the body as well as a major spiritual change. It may purify your personality, removing toxic patterns and habits that hold you back from success.

Rebirth can be quite messy. We are often put through a difficult process as we await the moment when we will be changed and ready to step into our new lives. But this process can be taxing and you may have to work harder than you thought to reach your goals. This bath can help soothe your soul and nourish your tired body as you move through the energies of rebirth.

For this ritual, you will need:

- Milk (coconut milk is a good vegan substitute)
- White rose petals
- 3 small pinches of motherwort

- Water charged with the symbol of the Vesica Pisces

- Pomegranate cut into quarters

Mary or sea bean, which has been used to ease the pain of child-birth, can also be used, as well as hag or holy stones, which were often used in charms to help ease the baby through the birth canal. You can add spring flower petals to the bath to represent new birth and the beauty of springtime, or hold them and use them as focal points during your bath. Water charged with the energy of the Chariot and Sun cards from a tarot deck can also enhance the power of this bath.

Begin by cleansing your body using your favorite method. Then fill up the tub with water and add the milk. This symbolizes the gift of nourishment from the mother. When the tub is full, sprinkle the white rose petals sunwise into the bath water, then do the same with the motherwort. Place your hands so they are touching the surface of the water and begin to charge it with your intent. Visualize the motherwort acting like stitches and pulling the rose petals together to form your intent in your mind's eye. Let this form morph into your rebirthed and whole self. Look at it lovingly and know that you are on the verge of being reborn to a greater purpose, and that what you are experiencing right now is temporary. You will soon be born anew. Allow this image to melt into the water, symbolizing creation, femininity, and birth.

Get into the tub, sit back, and relax. Allow yourself to process all that you have gone through, and love yourself for where you are. Anytime you come across a thought about who you were–a doubt, a regret, or a negative feeling–squeeze the blood (juice) from one of the pomegranate quarters over your body. Acknowledge your pain, your loss, and your grief for a moment. Then remember your strength and realize that you, like Persephone, will triumph over your darkness and arise from the underworld and into your new life—reborn different, but whole and ready to take on life gracefully.

Do this until you feel the ritual is complete, then drain the tub and stand up, letting the bath water run off your body and down the drain. This is a messy bath, so you will probably need to throw some seeds and other plant matter in the trash and rinse off in the shower.

Exercise: Sea Bath for Healing

Bathing in the sea is a beautiful practice for which you need nothing but your body and the ocean. You can bring a beautiful ritual vessel like a shell or a bowl to make it easier to control where the water flows. Begin by walking into the water far enough so that, when you sit down, it is up to your waist. You don't want to ruin the experience by getting knocked over by a wave, so sit in a shallow location where you are safe and can enjoy the water. Warm oceans in the summer are ideal for this particular ritual, but you can do it at any time. Just be sure to care for your own sacred vessel—your body.

When you are ready, begin to watch the waves and just enjoy the ocean's general splendor. After a time, choose a particularly large wave and begin to count the next nine consecutive waves. On the ninth wave, catch some of its water with your shell, your sacred vessel, or your cupped hands. Speak a short heartfelt prayer asking the ocean mother to lend you healing energies.

Then pour the water over your head. You can do this as many times as you like, but I recommend at least three or nine repetitions for a very powerful nine-waves healing bath!

Exercise: River Bath for Cleansing

In the spring, when the water is warm enough to sit in, wade out a few feet into a river and sit with the water up to your waist. Sit with your back facing the current so the water splashes on your back and flows past you. Allow the river to gather your unwanted energies, anxieties, and worries and cleanse you of them. Let them flow away from you and down the river.

Exercise: Protection Bath

Taking a bath prepared with sea salt is a great way to add protection to your weekly or monthly routine. You can add any other ingredients you feel are necessary, but the base ingredient here is the salt that you use. As you fill your tub, sprinkle the salt around it three times, casting away all negativity by saying:

Three times this salt is sprinkled about; I hereby cast all evil out!

Step into the tub and lead yourself through a visualization in which you see yourself protected by water and doing your work as a priestess without hindrance from outside forces. You can draw on a little plant-spirit help by adding hawthorn, rose, vervain, hyssop, rosemary, St. John's wort.

Exercise: Purification Bath

Purification is an interesting concept because, in many cultures and religions, it has become a toxic word that implies virginity or other ideas of perfection. In these cultures, purification has come to mean removing something in order to make you holy, divine, or desired. In the context of the water priestess, however, the word has no such connotations. Rather, it means that you must be pure of heart and mind, not in physical form or chastity. To be a pure water priestess, you must be true of heart, true of mind, and free from the bonds that hold you back. You must be free of forces that make you different from or less than what you can be. You must be the best version of yourself, without restrictions or chains.

Of course, there is an astral or spiritual aspect to this as well. Throughout life, we pick up energies as our experiences leave their energetic imprint on us. Sometimes these are negative energies and spirits–what I call astral nasties–that attach themselves to us without our knowledge. It is important to understand this, even if you are sure you don't have any negative attachments, because, as a water priestess, you may find that these nasties have attached themselves to those you work with or those for whom you have performed healing rites. You can use this bath to purify yourself of these negative energies, or you can modify it for others to use.

For this bath, you will need:

- 3 drops of cedar essential oil (optional)
- A pitcher, chalice, or sacred bowl
- Dried rosemary, hyssop, and white sandalwood
- Sea salt
- White candles

- Quartz crystals

- White rose petals delicately removed from the stem

Mix the rosemary, hyssop, and white sandalwood together and place a large handful of the mixture in a bowl. Pour several cups of boiling water over them and let the mixture steep until you are ready. Strain out the herbs and put the herbal water into a chalice, ritual bowl, pitcher, or jar.

Prepare your bath by cleaning the tub, then burn a mix of herbs and resins to purify the air. I suggest white sandalwood, myrrh, and rosemary. Allow the smoke to fill the room, then set the room with candles, crystals, and anything else of your choosing.

Add nine pinches of sea salt to the water and stir it sunwise nine times with your dominant hand, saying:

Nine times the waters turn,
Nine pinches of salt that churn;
Purification imbued in turn.

When you are ready, place the white rose petals and the herbal water where you can reach them and step into the tub. Do not sit down yet! Hold the rose petals over your head, tip your head back slightly, and pour the petals over your head, allowing them to fall across your chest, your breasts, and your tummy before reaching the water. As you do this, say:

With the power of the sacred rose that falls upon me,
I purify myself three times three.
With the power of my mind,

I break away the chains that bind.

Raise the container of herbal water–which should be warm but not hot (you want to purify yourself, not be scalded!)–over your head, letting it flow over you in three large portions, using about 1/3 of the water each time. With each portion, say:

Power of these herbs and water divine,
Wash over all that is me and mine.
Purify my mind, body, and heart
So that each day I have a brilliant start.
My soul and aura made anew,
No evil or bane can break through.

Be sure to empty all the water as the last portion pours out over your head and body into the bath. Make sure you use a lot of white rose petals. This bath is very powerful emotionally and spiritually, and the more rose petals you use, the better. Besides, it is pretty luscious!

Exercise: Heart-Healing Bath

For this bath, you will need a heart-shaped rose quartz and water from a sacred spring. If you want to call upon the energies of Aphrodite, you may want to use a statue or image of her. Place the heart-shaped rose quartz in a beautiful bowl and pour the sacred water over it. Use this water as the base for your ritual. You can also set pink rose petals and blue candles around the bath. Carve hearts into the blue candles, since that is what you are healing.

Make a powder of dried white rose petals, hawthorn flowers, rosemary, chamomile, hyssop, and a few slivers of vanilla by grinding them up with a mortar and pestle. (If you are substituting vanilla extract or essential oil, you will add it later.) First add the hawthorn flowers, asking them to heal your broken heart and allow energy to flow through it with ease and grace. Then add the white rose petals, asking them to purify all the negative feelings and energies associated with your broken heart. Add the rosemary, asking it for protection of your heart and for healing energies, then the chamomile, asking it to soothe your pain and nourish future love. Finally, add the hyssop, asking it to purify your heart and to allow healing to enter. Grind this mixture into a powder or chunkier herbal mix. Then add the vanilla slivers and charge the mixture with the heart-shaped rose quartz until it is ready to be added to the bath.

Fill your tub with water and add the herbal powder. As you do this, visualize your heart healing and moving to a better energetic place. If you are using vanilla extract or essential oil rather than slivers, you can add it directly to the bath here. Add the pink rose petals and kneel before the tub with your heart-shaped rose quartz in hand. Focus your intent and pray to your goddess, reciting this incantation:

Sacred waters pure and clear,
Bring in healing, draw it near;
Healing waters of love and grace,
Bless me and this sacred space.
Waters of transformation,
Swirl from deep below;
Mend this broken heart
With your sacred flow.

Then step into the tub. Hold the heart-shaped rose quartz and enjoy the luscious bath while meditating on how your heart was broken and how you can heal it. Then visualize sewing yourself back together, stitching up the wounds with a beautiful silver thread of divine light. When you are finished, step out of the tub. Place a few fresh petals and the quartz in a dish and allow the water to drain. Remove the petals left in the tub. If you have an outdoor shrine, sprinkle them there.

Finish the ritual by taking the dish of rose petals and rose quartz to your shrine. Wrap the crystal in silver thread (metallic needlepoint floss works well) and keep it on your shrine. When you feel as if you may be slipping, return to your altar, hold the crystal, and recite the incantation again. Remember: Healing takes time, so be kind to yourself and invite grace in.

Exercise: Bath for Forgiveness

This bath is fun, yet challenging. For it, you will need either freshly gathered snow or crushed ice in a bowl, as well as white rose petals in another bowl and a significant amount of dried hyssop ground to your liking.

While you are gathering your snow or ice, think of the person or act you need to forgive. You can forgive yourself, or perhaps someone who has wronged you, left you, or hurt you. Set the bowl of ice next to the tub, but do not put it in the water yet. Keep the white rose petals in a bowl beside the tub as well. Fill the tub, making sure that the water is warm but not too hot, then sprinkle the hyssop into the water and ask for its spirit to help you relax and release your frozen heart.

Step into the tub and relax. Bring yourself to a light meditative state and think about the person or action you want to forgive. Let your mind wander through what led you to this place and consider what may be stopping you from moving forward. Then empty the bowl of ice onto your chest and focus on how much it stings—how the hard and cold energies actually hurt you. As the ice melts into the warm water, focus on the energies of transformation and the warming and softening of the ice. Visualize that which you need to forgive as the ice melts. Allow your own heart to soften and release the hurt.

Then empty the bowl of white rose petals onto your chest, where the ice was. Notice how different they feel. Notice how soft and lovely they are and how good they feel on your skin. Spend time playing with the petals, holding them, feeling them. As you do, invite grace in. Allow your heart to heal, to forgive that which caused you harm. Sit back and relax, enjoy the rose petals, and purge yourself of any negative energies. When you are done, drain the water and sprinkle the white rose petals outside or at your shrine.

Exercise: Boundary Protection Bath

For this bath, you will need several rose thorns and/or hawthorn thorns. Avoid using blackberry or raspberry thorns, as they are harder to work with. Hawthorn is the plant of both sex and death—a plant of the liminal realms that is found in hedges across the landscape of Britain, Scotland, Ireland, and Wales. Roses are also well known as plants that create boundaries. The beauty of their scent and their delicate petals represents the very sacredness and soul of the priestess. But remember the thorns! Gather the thorns you need directly from the plant if you can, but be sure to approach the plant with reverence and perhaps a little offering of sacred water, a quartz pebble, or a shell. Place your hand on the plant or in the dirt directly under it and send your energy into the earth to connect with the plant on a psychic level.

Tell the plant that you are harvesting a few of its thorns to use in ritual, then wait and allow your own intuition to guide you. Feel out the energy of the plant. If it is negative, thank the plant, give it your offering, and move on to another one. If it is positive, thank it, give it your offering, and gently remove a few thorns. If you happen to get pricked in the process, consider your blood as an exchange of life force with the plant itself. You are taking part of its body; it is taking a part of yours. If this happens, stop and spend a few moments in meditation to see if the plant is willing to work with you and bond with you as a plant-spirit ally. Take some petals with you to dry. You can return before your bath to gather fresh petals if you like. Allow both the thorns and petals to dry.

You can also incorporate hawthorn leaves or berries into your bath. No matter which part of the plant you decide to use, allow it all to dry so that it can be ground into a powder. Be sure you have a significant number of thorns in mixture. You will need a strong mortar and pestle to grind them. I prefer the type made of pumice stone because it is rough and grinds the harder plant bits more easily. Sift any of the larger pieces out of your powder. Then gather your fresh rose petals and sea salt.

Fill the tub all the way up. As you do, sprinkle the sea salt into the water in a sunwise direction while saying a blessing prayer of your choice. Then add the fresh rose petals. When you are ready, hold the bowl containing your ground powder in both hands and visualize thorny vines growing out of the powder and into the tub. Sprinkle the powder on the water in a sunwise direction and step into the tub. As you sink into the luscious petals, visualize the vines growing all around you, gently enclosing your aura and energetic space while protecting you from outside forces with a thorny barrier.

Exercise: Purification Baths for Others

You can create purification baths for others by using the same instructions I have given above and adding any wisdom you may have learned from your practice to help those you are healing. Then give them some of the herbal water that you prepared for yourself, as well as a vial of sacred water to pour into the bath. When you prepare the water for others, you may need to add a bit of vodka to preserve it, as it does contain herbal particles that can molder after a few days.

Grounding

Grounding is an essential part of every magical practitioner's work. Whether you are leading someone through a healing ritual or using it for yourself, it is essential and very beneficial to be grounded. Water is especially useful in grounding, as it allows for varying degrees of power. Water flows, so its powers are malleable. When you are in need of a light grounding meditation, you can modify the flow of the water to generate a more gentle grounding and cleansing energy. Or you can speed up the flow and increase the power if you are in need of deep grounding.

Exercise: Waterfall Grounding Meditation

Begin by finding a comfortable position. This usually means sitting, but lying down and standing can also work, depending on the situation. Take a few deep breaths and allow your mind to quiet. As thoughts come to your mind, acknowledge their presence and allow them to leave just as quickly as them came. Bring your mind to stillness.

Begin to visualize a waterfall. As you do this, see yourself walking under the waterfall so that it flows gently over your head. See the water begin to mingle with your own energy. As it does, it brightens and enhances your energies where they are already light and beautiful, while the flowing water gathers and absorbs the negative and sluggish, or anxious and baneful, energies and washes them off your body and out of your energy field into the earth. Slow down the flow for a gentle grounding; speed it up for more power. When you feel as if you are cleansed, gather the water in your hands and visualize yourself saying:

Water, cleanse and ground me today; wash all this bane away.

Allow your mind to fade back to darkness and take a few deep breaths to come back to this time and space.

Exercise: Water Grounding Ritual

For this ritual, you need a glass or small bowl–nothing large, just something big enough to hold some water. You can use ordinary tap water or you can use sacred water. You can also prepare a grounding gem elixir with smoky quartz (see chapter 4).

Hold your small dish or cup of water in your hands and focus on all that is making you anxious, flustered, or sluggish–all that needs to be removed from your energy so that you can feel grounded and ready for ritual. Once you have identified these things, gather your thoughts and take a deep breath, then focus and blow your energy out through your lips and into the water. You should see ripples on the surface. Do this again and again until you feel grounded, then dump the water out.

If you are doing this ritual with others, have a larger bowl handy into which the water can be dumped. Then dispose of the water after the ritual as part of your clean-up.

Writing Water Rituals

Writing rituals is a fun and worthwhile task. You can create your first ritual using the following outline and then use that ritual as a foundation for your next, and then your next, until you have developed your own method and way of doing things. At a certain point, you will not need to follow an outline and your rituals will just flow through you. By then, you will be able to recognize and work with any roadblocks that may pop up.

Begin by answering these questions:

- What is the main purpose of my ritual?

- How do I want to express this?

- Who is this ritual for?

- How many people do I hope/expect to attend? Do participants need to bring anything?

- What items will I need for the ritual? How will I use them?

- What type of water do I need to gather or prepare for this ritual?

- Will I be using candles, paper, flowers, shells, crystals? How will I use them?

- Will I be crafting an altar and what do I want to put there?

- What can I bring to make participants more comfortable?

- What issues could arise and create potential conflict? How will I handle them?

- How will I create sacred space?

- How will I ground participants?

- Will participants play a role in the ritual? Will they chant or charge items? Share their feelings or stories? Create magical charms? Charge water or raise energy?

By answering these questions, you will gain an idea of how to begin framing your ritual. You may find that you want to include certain items, but that they just don't fit into the ritual. Try not to be too connected to any particular outcome and let the theme flow through you, guided by the water.

Start by writing an outline—perhaps something like this:

- Welcome participants.

- Invite everyone to be seated.

- Greet everyone with watery words of wisdom.

- Explain the ritual. This is really important, as it may make people who are dealing with anxiety more comfortable if they know the details of what they will be required to do beforehand.

- Hand out items to used in the ritual.

- Invite participants to set intentions or teach them any chants or hand gestures they need to know.

Once you have the outline of your ritual clear in your mind, you are ready to begin.

Invite participants to take a few deep breaths (three or nine) to connect with water and move into a meditative space. Lead them through the grounding meditation on page 209. Begin by passing out any supplies and reminding everyone what will happen next, then move into the main portion of the ritual, following your outline.

When you are finished, bring everyone's attention back to the circle and allow time for people to process or share their experiences if it is appropriate. Pass a small bowl of blessed water around and let participants anoint themselves with it to ground them back to this time and space, and to release the energies you were just working on.

Thank everyone for joining you, let them know you will be available if anyone needs to ask questions, and then leave them with words of wisdom–perhaps a type of homework through which they can continue to contemplate the work they have done.

Chapter 9

ON THE PRIESTESS PATH

$\mathcal{T}$he way of the water priestess is ancient, complex, and richly rewarding. Before embarking on this path, however, you must first define yourself and then decide where the journey will lead you. You do the first by creating a sacred symbol or sigil that will act as your magical name. You do the second by identifying your strengths and allowing them to help chart your course.

Creating Your Symbol

Your priestess symbol acts as a representation of you. It can also act as a magical seal or signature with which you can sign your work. You can use it to signal the beginning or ending of work at the water's edge by drawing it in the sand. This sacred symbol will become a powerful representation of you and your work.

But first you must create and charge it, and then bind it to both you and the water.

There are many ways to create your own priestess symbol. Many people use magical alphabets or words from their cultural background. You can also use shapes and symbols that are already significant for you and then combine them together. You can overlap shapes and rearrange them, turn them over, or remove parts of them. Do whatever feels right for you until your symbol carries a complete representation of you and your work as a water priestess. The important thing to remember is that, no matter what you choose, the symbol must be unique to you. It is not intended to be a popular or universal symbol, although you may want to incorporate them into your design.

Next, you must imbue this symbol with your energy. Your thoughts and intent have already put energy into it, and the creation of it on paper has also energized it. But now you must awaken its power. To do this, place the paper with the symbols on it between your hands. Close your eyes and focus on the symbol. See it in your mind's eye. Visualize it pulsing and glowing with a watery blue light. See it grow and begin to merge with your heart center and into your body. Repeat this with your Third Eye and your belly area.

Exercise: Charging Your Priestess Symbol

When your symbol is ready, place it in on your working surface and set a completely clear bowl directly over it, so you can see the symbol through the glass. Fill the bowl with any water you choose, making sure that you use water with which you have a deep connection. Keep in mind that you will be taking this water to your local water source and pouring it out, so be sure it is clean drinkable water without additives. If you can, use water from the original source, but sacred water from a spring is also a great choice.

Leave your bowl and symbol to sit overnight. The next day, kneel in front of your altar dressed in your ritual garb. Dip your fingers into the water and draw your symbol on your Third Eye, then dip again and draw it onto your heart center. Dip a third time and draw the symbol on your belly area. You can also use your hands and feet as alternatives here. Just make sure that you adjust your visualization accordingly.

Once you are done, visualize the water symbol again, this time on the surface of the water. When you have it fixed in your mind's eye, visualize the blue watery symbol rising up out of the water and fusing with your Third Eye, then with your heart center, and finally with your pelvic area. Spend more time listening to the water and fusing your energy with the symbol and the water. When you are finished, add a taglock to the water–a single hair or a small amount of saliva–and pour it into the nearest water source. While you do this, whisper to the waters that this is a representation of you and your work with the water, and invite the water spirits to recognize it as such as well.

Defining Your Path

I hope you have found ideas in this book that have deeply inspired you–ideas you can incorporate into your own path to bring it into closer alignment with water. You may even have found ideas that you don't like! These are all important to keep in mind and to evaluate as you move forward on your path as a water priestess. Once on this journey, you can focus your work on practices at which you excel and on those that you deeply enjoy.

The way of the water priestess is as unique for each of us as a water crystal—something that belongs to water, but has its own unique structure. While you may be aligned in some ways with other water priestesses, your own path and practice may be very different from theirs. Even if your practice seems similar, it will still be unique and reflect your own strengths and preferences as a priestess. So don't be upset or frustrated if you see another priestess offering services similar to yours. The world is so in need of our craft right now—so much water to be healed and so many people to be helped—that we could double, triple, or even quadruple our numbers and our work would still not be complete.

Although there are many ways to serve the water, as a water priestess, you must narrow down your offerings so that your practices are sustainable and not overwhelming. Remember: You can't pour from an empty cup. There is no possible way that you can embody every single aspect or practice of a water priestess and still pay the rent, feed your children, and hold down a job. If you have committed to working as a full-time water priestess, you still must narrow your focus and specialize so you can bring the full power of your practice to your efforts to help the water and water spirits.

Someone once asked me what my specialty was as a dancer. I told him I was pretty good at many types and styles of dance. He laughed and said: "You can't be great at everything." Then he went on to explain that I should pick one or two types at which I wanted to excel, rather than using all my energy trying to be "pretty good" at a variety of them. His advice really stuck with me and, as my life progressed, I found this wisdom applied in many ways to my water practice as well. When we try to do everything, the quality of what we do suffers from lack of time, dedication, and energy. When we narrow our focus to a few things, we can become great at those things, and can then offer that greatness to others. So learn to focus your work on your natural abilities rather than on those things you struggle to practice or learn.

In fact, it is counter-intuitive to try to become great at something that doesn't sing to your soul. Don't try to shove yourself into a box that doesn't fit your personality, your natural alignment, or your bliss. Water is liquid; it wants to move. If you try to contain it in a vessel that is not quite right, it will work hard to find a way to escape—whether through energy leakage or just by finding the cracks in the vessel. Do not limit yourself for others; do not force yourself into stagnation; and do not focus on things that don't bring you joy!

It is important that you keep your energy clear and moving. When we do the things we are "supposed to do" rather than the things that are our divine bliss, we put ourselves through unnecessary suffering that leads to blockages. So be sure that the practices you choose to be part of your path are the things that nourish you, so that you can nourish others in turn.

Here are a few questions you can ask yourself to help you narrow your focus:

- When I am in ritual, what do I like the most?

- When I am in ritual, what do I like the least?

- Do I like working with others? If so, do I like working within a circle? In rituals?

- Do I prefer solitary work?

- What element do I align with most? How can I incorporate it into my practice?

- What element do I align with least? How will that inform my practice?

- When I think about water priestesses, what types of activities and services do I see them offering? Are these things that make me feel good and that are in alignment with my energy? If not, how can I restructure my vision of a water priestess so that it fits the things to which I am drawn?

- What things do I not like? What things am I not good at? Is it fear that keeps me from these things, or are they naturally out of alignment with my path? If it is fear that is blocking me, is there a way that I can overcome it and integrate them into my path?

- What are the three most inspirational things about the way of the water priestess I have read, seen, or experienced? How can I integrate them into my own practice? How can I create water rituals with these types of practices?

- What do I think is the most important thing about water priestess work and how can I accomplish this?

- What tools do I like using? How can I incorporate them into my work?

- What type of water am I drawn to and how can this inform my work and rituals?

- Do I prefer to work at my home temple or at the water's edge?

- What other modalities am I already trained in? What modalities do I want to train in? How can they support or enhance my work as a water priestess?

- Do I consider myself a healer? Warrior? Facilitator? Oracle? Other type of worker? How can this act as the central focus of my path?

Once you have answered these questions, you will be ready to define your path as a water priestess and dedicate yourself to it.

Dedicating Yourself to the Path

The following ritual is one you can use to dedicate yourself to your chosen path as a water priestess. There is no such thing as a self-initiation; you cannot initiate yourself into mysteries that do not belong to you. But you can dedicate your life and your path to the waters and to serving them in some capacity.

There is a chance that this type of ritual may trigger a spirit initiation or a death-and-rebirth cycle. If the water accepts your dedication and your words have deep meaning and significance to you, be sure you are ready when you say: "May these sacred waters wash away all that does not serve me." While this ritual may remove old memories, addictions, and negative or counter-intuitive habits that stop you from being your best self, sometimes the things that do not serve you are things that you love or don't want to lose. Just be sure you are ready for it or change the words accordingly.

You have probably already spent quite a bit of time in dedication, even before you picked up this book. But your path is important, and marking the point on your path at which you decide—or perhaps it is decided for you—that you will live in service to the aquatic realms is pivotal. Even if you started on this journey long ago, you can do this ritual on your priestess anniversary, on a sacred day like Imbolc, Beltane, or Samhain, or during a Full Moon, or even on your own birthday to re-dedicate yourself.

The setting for this ritual matters. If you have already connected with a body of water and want to associate your dedication to that particular shore, you can perform it there. Or you can do it in your home temple space if you already have one set up. If you choose to do an on-site ritual, scope out the area first and choose a time when you can be private and undisturbed. If you want to focus your work more on the sea, work on your favorite beach and change the words in the ritual to reflect that. You can also combine this ritual with a sacred pilgrimage, as described on pages 228-229. But if you choose a place near your home, you will be able to tend the waters where you were dedicated more easily. No matter your choice, this is a beautiful and special rite!

This ritual is about becoming a well maiden, a water bearer, a sea priestess, a sacred water woman. It is about dedicating yourself to water, or to a specific body of water. It marks a turning point on your journey at which your life becomes focused on water–healing it, tending it, holding its secrets, protecting it, serving it, and using its magical powers to heal others, the land, and this world. In it, you will use many of the tools you have learned about in this book. You will part the mist, create the nine sacred waters blend, take a cleansing bath, purify the temple, etc. You may also want to decorate your temple space for this ritual, even if you decide to perform it at the water's edge. You can mark the occasion with fancy altar cloths, favorite flowers, fresh offerings, or special incense.

Exercise: Water Priestess Dedication Ritual

Dress in your sacred robes and special jewelry. Select these carefully, as they will be a representation of your priestess work and will become a taglock that connects you to your dedication and the work you will do as a water priestess.

Before you begin, review the previous chapters and be sure that your temple is properly built and cleansed. Then perform the purification ritual bath, create sacred water, and gather the necessary tools.

For this ritual, you will need:

- A representation of each element having a connection to water

- Cleansed altar space

- Any sacred jewelry or robes you want to dedicate to your path and your work

- Nine sacred waters blend (see chapter 8)

- Sacred vessel

Spend time cleaning your space and be sure that the temple's energy is cleansed and ready. To begin, follow these steps, then move directly into the ritual.

- Clean your temple on the mundane level, then cleanse it energetically.

- Decorate your temple, shrine, and altar for this special occasion.

- Take a purification bath (see chapter 8).

- Set the ritual with your most beautiful altar cloths, your most important tools, your favorite flowers, etc. Gather all your materials and place them where they should be located. Be sure to include your altar, sacred vessel, sacred water, and offerings, as well as a representation of each element in its proper position.

- In the north, place a dish of sand or sea salt, or shells or other Earth/Water representations.

- In the east, place incense in shells, feathers from aquatic birds, and other airy things that align with Water and Air.

- In the south, place candles floating in water, or candles anointed with sacred water.

- In the west, place any representation of Water or a beautiful jar of sacred water from your favorite source.

Begin the ritual by standing in the west. Pick up the water and say:

Powers of Water in Water, I seek your presence here today; please wash away all unwanted energies.

Walk around the circle sunwise and sprinkle the water as you go, saying:

I consecrate and purify this sacred space with the power of Water in Water.

When you arrive back in the west, put down the water and move on to the north. Pick up the sand or salt, and say:

Powers of Earth in Water, I seek your presence here today; please ground this space and all the energies present.

Walk around the circle sunwise while sprinkling the sand or salt, saying:

I consecrate and purify this sacred space with the powers of Earth in Water.

When you arrive back in the north, put down the Earth representation and move on to the east. Pick up the Air representation and say:

Powers of Air in Water, I seek your presence here today; please blow away all unwanted energies.

Walk around the circle with your incense sunwise and allow the smoke to fill the air as you go. As you walk, say:

I consecrate and purify this sacred space with the power of Air in Water.

When you arrive back in the east, put down the Air representation and move on to the south. Pick up the candle and say:

Powers of Fire in Water, I seek your presence here today; please burn away all unwanted energies.

Walk around the circle sunwise, carrying the candle, and say:

I consecrate and purify this sacred space with the power of Fire in Water.

When you arrive back at the starting point in the west, put down the candle.

Once the circle has been consecrated and purified, you need to protect the space. To cast a circle of protection, enact these steps in both the mundane and spirit realms through visualization. Start in the west and walk sunwise around the circle three times, each time chanting the protection incantation on page 225. If you feel you need more protection, you can do this six or nine times. As you circle, be sure to visualize a protective bubble forming above, around, and below you. Remember: This a sphere, not a circle!

Water spirits from down below,
I ask for your protective flow.
Water spirits from up above,
I ask for your protective love.
Water sprits on this ground,
I ask for your protective powers all around.

Repeat this at least three times. On the final repetition, add:

Water spirits who draw near,
Protect this space I hold dear.
With grace and love
From here, below, and above!
So mote it be!

Begin the next portion of the ritual by parting the mist (see chapter 2) and moving into an altered state. You can skip the tea and just do the visualization if you prefer. Then stand, kneel, or sit in front of your water altar and breathe rhythmically until you are comfortable, open, and ready to begin. You can call upon any water goddesses, spirits, or guides to be present to witness your rite.

Pour the nine sacred waters blend into your sacred vessel (see chapter 8). If you are working with rose petals, you can add a few here. Hold your hands over the vessel and visualize your priestess self. Hold that image in your mind. See yourself doing the work you have been drawn to do; see yourself in your robes and sacred adornments. Then say:

Today I dedicate myself to this path.
May all the water spirits who witness this rite
Bless me and my work tonight.
Guide me through my spirit flight.
Bring me the energies the water holds dear
So that I may do my work with no fear.

Dip your finger into the bowl and anoint your Third Eye with the symbol of a crescent Moon. Say:

I dedicate myself, my path, and my life
To the path of water, through both blessings and strife.
May my eyes be open to the work that is needed
And may my intuition steer me true.

Dip your finger in the water again and draw a down-pointing triangle on your heart center. Say:

May each and every day be in service to the water,
May my love for the water and spirits that dwell within guide me on
this path,
Through love and grace.

Dip your hands into the water and place them together as if in prayer. Then slowly move them down, opening them so they form a down-pointing triangle with both thumbs and both index fingers touching. Hold your wet hands in this triangle over you belly, and say:

I dedicate myself to the spirit of water,
May you wash away that which does not serve me
So that I may serve the water to my fullest capacity.

Dip your finger in the water again and draw a spiral on the back of your left hand, then again on your right. Hold your hands with your palms together, and say:

May all that these hands do be of service to the water.

Do this on the tops of both your feet, and say:

May every step I take be in service to the water;
May my feet steer me true,
So that I can walk the path of water with ease and grace.

Sit back and relax and spend time in meditation. Then record any visions, thoughts, symbols, or ideas you may have experienced in your Book of Tides. Take a vision journey to your astral water temple (see chapter 4) and perform this same rite there. Dip your sacred jewelry and any tools you want to dedicate to your path in this water, blessing them and asking the water spirits for blessings as well.

When the ritual is finished, pick up the sacred vessel holding the water and charge it one more time with positive dedicative watery energy. Then say:

Water spirits both far and near,
Bless this body and all it holds dear.
Remove my fears and blocks
So that I may honor you on my walk.
I ask for your blessings, your wisdom and grace;
Bless your priestess in this space.

Pour the water from the sacred vessel over your head and allow it to pour down your body. Spend more time in contemplation.

When you are ready to end the ritual, walk to the west and say:

Powers of Water in Water, thank you for your guidance and protection today.
As you go back to your watery realms, I thank you.

Move to the south and say:

Powers of Fire in Water, I thank you for your guidance and protection today.
As you go back to your fiery realms, I thank you.

Extinguish any candles, then move to the east and say:

Powers of Air in Water, I thank you for your guidance and protection today.
As you go back to your airy realms, I thank you.

Be sure to extinguish any incense that is still burning, then move to the north and say:

Powers of Earth in Water, I thank you for your guidance and protection today.
As you go back to your earthly realms, I thank you.

Remember the date on which you perform this ritual and consider repeating it every year in a slightly different version to re-dedicate yourself to your path and renew your vows to the water and the water goddess.

The Sacred Pilgrimage

Pilgrimage has been a common practice for spiritual and religious seekers for hundreds, if not thousands, of years. Various holy lands and sacred landscapes have been visited by the faithful all over the world throughout time. Even today, places like Jerusalem, Mecca, Nepal, Tibet, Glastonbury, Kildare, and even Sedona are commonly visited for religious and spiritual reasons. Throughout history, pilgrimages to holy wells and sacred lochs were common. Consider taking your own pilgrimage–perhaps to a sacred water site in your local area, or even a distant voyage across the water.

This kind of journey is sometimes called an *immram,* a hero's journey across the water and into the Otherworld. It may seem strange to you that a pilgrimage in the mundane world can also create a connection to the Otherworld. But for priestesses and magical people, the two are not so different–especially if the journey is of a magical, mystical, or spiritual nature.

Once you have planned your pilgrimage, consider what you have learned here about making offerings, bottling water, and other priestess practices. When you arrive at your destination, be very mindful of what is best for the spirit of the place–the *genii loci.* You can perform the dedication ritual described previously anywhere–at a distant location, or near your home at your local water source.

CONCLUSION

Water's wisdom, like her nature, is vast and deep. She can deliver you lessons of light and joy, or of pain and sorrow. She is the great spiral, always spinning and always flowing. She can teach us to be like ice and stay still, to freeze in our tracks and protect our energies. Or she can teach us how to warm a heart, how to release petrified energy, and how to dance in the wind like beautiful snowflakes. She is the maiden, the mother, the crone, the warrior, and sacred sexuality. She teaches us the nature of destruction, but also the nature of life.

The priestesses of the ancient temples were guided by the old gods and had direct contact with them. They never died; they never went away. They still dwell here within the water, waiting for water priestesses to awaken and rise again. With their help, we can restore this wasteland.

I have a vision that, in my lifetime, before I pass into the realms of our ancestor priestesses, there will be 9,000 working water priestesses worldwide, with temples, healing centers, community rituals, and anything else that will help us come together and change this planet. We must work to revive the old ways and bring a global sacredness back to water. We have been given a task—an immense task—to return the waters to a sacred nature. I hope this book has inspired you to work along with us. It is time to return to the temple—to the water—and to work for good in any ways we can.

In these pages, I have called forth a sacred remembrance that I offer to all those who are drawn to the sacred service of water. I charge you to awaken and remember the ways of the ancient water priestesses and welcome them into your own path and practice. Share the wonders of the sacred waters–from river to river and from sea to sea. Arise and step into your power in this moment, in this sacred hour.

I leave you with this parting prayer.

Prayer for Water Wisdom

Goddess of water, hear my prayer;
Grant me wisdom to dare.
Goddess of water, hear my plea;
Grant me wisdom to be ever flowing free.
Goddess of water, please teach me your ways,
How to ebb and flow, and Moon knowledge in each phase.
Goddess of water, bless my sacred work;
Give me deep knowledge to overcome each trial,
To prevail in the storm and heal the water mile by mile.
Goddess of water, please grant me your sight;
Ancient wisdom I do gain.
I remember the old ways and honor them again;
Wisdom from water and second sight,
Strengthen my wisdom with water's might.
To view visions on your mirrored surface by the candle light.
Lend your wisdom drawn down to impose;
Fill my cup with the sacred feminine rose.

Wisdom of the chalice I ask you to impart;
Deepen my connection to the waters and heart.
But then so mote it be!

INDEX OF RITUALS AND PRAYERS

Chapter 4: Raising the Temple

Chapter 5: The Healing Power of Water

Chapter 6: The Wisdom of Time and Tides

Chapter 7: Water Divination

Chapter 8: The Art of Ritual

Chapter 9: On the Priestess Path

Conclusion

ABOUT THE AUTHOR

Annwyn Avalon, a water priestess and water witch, is the founder of Triskele Rose Witchcraft, an Avalonian tradition. She has devoted her life to the service of water and the water spirits in numerous ways including the study of art, witchcraft, magic, and priestess arts. Annwyn is an initiated priestess, Reiki master teacher, and award-winning dancer. She has a BFA in sculpture and a BA in anthropology with emphasis on plant and human interactions. Annwyn has also received an apprentice certificate in herbalism. She writes for *Patheos* under the name the Water Witch as well as for *Magical Times* magazine in the UK and has contributed to other published works including *This Witch Magazine* and *The New Aradia: A Witches Handbook for Resistance*. The author of *Water Witchcraft: Magic and Lore from the Celtic Tradition*, Annwyn lives in rural Oregon with her dogs, cats, chickens, and a peacock or two. She spends her days in the garden and her nights dancing in various productions throughout Portland, both as a soloist and with other sacred dancers. Annwyn provides water priestess training and other courses both locally and online.

Visit her at *www.WaterPriestess.com, www.WaterWitchcraft.com.* and *www.TriskeleRose.com.*

TO OUR READERS